Active Learning Workbook

for

Introductory Chemistry

An Active Learning Approach

Second Edition

Mark S. Cracolice
University of Montana

Edward I. Peters

Australia • Canada • Mexico • Singapore • Spain • United Kingdom • United States

Printed in the United States of America
1 2 3 4 5 6 7 07 06 05 04 03

Printer: Globus Printing
Cover image: Tony Stone/Paul Morrell

ISBN: 0-534-40681-5

For more information about our products, contact us at:
Thomson Learning Academic Resource Center
1-800-423-0563

For permission to use material from this text, contact us by:
Phone: 1-800-730-2214
Fax: 1-800-731-2215
Web: http://www.thomsonrights.com

Brooks/Cole—Thomson Learning
10 Davis Drive
Belmont, CA 94002-3098
USA

Asia
Thomson Learning
5 Shenton Way #01-01
UIC Building
Singapore 068808

Australia/New Zealand
Thomson Learning
102 Dodds Street
Southbank, Victoria 3006
Australia

Canada
Nelson
1120 Birchmount Road
Toronto, Ontario M1K 5G4
Canada

Europe/Middle East/South Africa
Thomson Learning
High Holborn House
50/51 Bedford Row
London WC1R 4LR
United Kingdom

Latin America
Thomson Learning
Seneca, 53
Colonia Polanco
11560 Mexico D.F.
Mexico

Spain/Portugal
Paraninfo
Calle/Magallanes, 25
28015 Madrid, Spain

Contents

Preface

To the Instructor

The Active Learning Workbook was written to accompany *Introductory Chemistry: An Active Learning Approach,* Second Edition, by Mark S. Cracolice and Edward I. Peters. The purpose of the workbook is to provide a supplement to the text to help students, who are most likely in their first college chemistry course, learn to be active learners of chemistry. We have found that many students try to study for a chemistry course by literally *reading* the textbook. We also find that many students "do" end-of-chapter questions, exercises, and problems by reading the question and then flipping to the answer section and reading the answer. We have provided this workbook as a tool to help you encourage your students to discover how to actively work at learning chemistry.

Both research and common sense tell us that active learners—those who work problems as they study—learn more than students who are passive in their approach to learning. George Polya, one of the most influential mathematicians of the 20th century, said, "Mathematics, you see, is not a spectator sport." This workbook is written with that idea in mind, as translated to introductory chemistry.

The Active Learning Workbook has chapters that parallel the textbook chapters. Each workbook chapter has two sections. First, there is a section of Target Checks and/or semiprogrammed Examples. These are designed to be completed immediately after each corresponding Target Check or Example in the textbook. A easily identifiable callout is given in the textbook at each point when a student should do the accompanying exercise in this workbook. Each exercise is designed with sufficient space for students to write their answers directly in the workbook. Target checks are answered at the end of each chapter. The semiprogrammed Examples are "answered" as the student works through the problem.

The second section of each chapter is a set of Questions, Exercises, and Problems that parallel those found at the end of each chapter in the textbook. The answers to these questions are not provided to the student. Therefore, they can be used for homework assignments. Space is provided after each question for student answers. The pages of *The Active Learning Workbook* are perforated so that they can be torn out. They are also three-hole-punched so that returned assignments can be stored in a standard binder. The *Instructor's Manual* gives simple answers and complete solutions so that you can post answers or solutions after the homework due date or provide these to graders as a key for correcting and evaluating the homework. A popular time-efficient strategy for homework assignments is to give a percentage of the total grade for effort (counting the number of assigned problems with a reasonable attempt at doing the work) and a percentage for correctness of a selected subset of the complete assignment.

If you choose not to use the workbook questions for homework, they can be used as an ungraded homework assignment for students. The workbook will help students organize their work.

We look forward to your suggestions about *The Active Learning Workbook.* Please contact us with your comments.

Mark S. Cracolice
Department of Chemistry
The University of Montana
Missoula, MT 59812
markc@selway.umt.edu

Preface

To the Student

The Active Learning Workbook was written to accompany *Introductory Chemistry: An Active Learning Approach,* Second Edition, by Mark S. Cracolice and Edward I. Peters. The workbook was written to help you learn chemistry *and* to help you develop good study habits that will serve you well in future science, mathematics, engineering, and technology courses. The key to success in courses like these is to engage in active learning. This workbook makes active learning easier by providing you with problems to work as you study the textbook and problems to solve after you complete a chapter, both in one convenient package, complete with spaces to answer our questions.

What, exactly, do we mean by *active learning*? Active learning is doing anything but passively listening to a lecture or passively reading a textbook in the same way you read a novel. Writing, thinking, discussing, and working problems are examples of things you do as an active learner. Since our job as authors of your textbook and workbook is to provide the textbook for your course, we will focus on active learning as it applies to what you do outside of class.

First consider the textbook itself. You can be an active learner when you study the book by taking notes as you read and working the Target Checks and Examples when you come to them. We've even provided space in the textbook for you to write your answers to our questions as you work the examples.

After almost every Target Check and Example, you will see an instruction that says, "Do Target Check or Example X.Y now." This instruction is a reference to this workbook. When you see this instruction, stop studying the textbook and do the assigned problem in this workbook. If you are satisfied that you understand the concept after completing the workbook problem, continue with your study of the textbook. If you are not confident in your understanding, go back and work through the textbook section again until you are satisfied.

The Questions, Exercises, and Problems in this workbook are similar to those found in the textbook. The textbook questions have complete solutions at the end of each chapter; the workbook questions are unanswered. Your instructor will choose how to deal with the solutions to workbook questions. Follow his or her instructions about the workbook questions.

Please take time to read and contemplate Chapter 1 of the workbook. It does not have anything to do with "chemistry," but it contains suggestions and ideas that may make the difference between succeeding and failing in your introductory chemistry course.

If you find any parts of this workbook to be factually inaccurate or unclear, please pass your comments along to your instructor, who will help clarify or correct the passage for you and also let us know about the problem so that we can correct it in future printings and editions.

The hard work that you put into being an active learner of chemistry will pay off in the long run, both in content knowledge and in improved thinking skills. We hope that the learning tools we've provided make chemistry a bit more understandable and enjoyable.

Chapter 1

Introduction to Chemistry

Introduction to Active Learning

Learning chemistry is difficult. There are no simple shortcuts to succeeding in a chemistry course. Whether or not you succeed in this introductory chemistry course will largely be determined by the choices you make now. Are you willing to make the commitment, in terms of self-discipline and time, to succeed in this course? Are you willing to defer short-term gratification for the long-term goal of learning chemistry? Are you willing to commit enough *time* to learning chemistry?

Let's look at five characteristics of students who are successful in chemistry courses and the corresponding characteristics of those who are not:

- Good students accept the challenge of learning chemistry and are willing to work hard. Poor students believe that chemistry is too difficult and procrastinate from the beginning of the course.

- Good students strive for complete understanding of each performance goal presented in the course. Poor students settle for something less.

- Good students study regularly. Poor students cram for exams.

- Good students work to develop better study skills. Poor students are satisfied with their present skills.

- Good students adapt to the demands of individual courses. Poor students study in the same way for every course.

How many of the characteristics of good students do you possess? Are you willing to try to develop your study skills? If so, we will outline some strategies in this chapter to help you succeed in your chemistry course.

Developing a Study Plan

As you begin this chemistry course, you should spend a few moments thinking about how you will go about studying for the course. We suggest a three-step study plan:

Step 1: Schedule your time

The beginning of developing a study plan is to decide how to commit time for studying. A good rule-of-thumb is to plan on about two hours outside of class for every hour in class. Thus a three-hour-a-week course, for example, would require six hours of out-of-class study each week. Ideally, you should schedule this time just as you schedule your classes. If your course lecture is three hours per week, then one hour per day, Monday through Saturday, will be a good starting point. If your schedule doesn't permit daily study, 2 hours per day, Monday, Wednesday, and Friday would be an alternative. However, avoid the tempting trap of studying all six hours on one day. That's a bit like deciding that taking one vitamin a day is a good idea and then implementing your plan by taking seven every Sunday. It's the *daily* dose that's critical. You need to schedule your time so that you have a study session sometime after each lecture and before the one following it.

Step 2: Determine what you are expected to learn

In general, most chemistry instructors expect you to know more than what is presented in lecture but less than the entire contents of the textbook. This leaves quite a wide range of possible expectations. There are usually many sources of information in a chemistry course. Common learning tools include the textbook, study guide, lecture, laboratory, demonstrations, handouts, homework assignments, supplementary books, computer programs, and audiovisual materials. Which of these are important in your course?

In most chemistry courses, the two primary sources of information are the lecture and the textbook. We will discuss these in more detail later in *Step 3*. For now, you need to assess the relative importance of the learning tools available to you.

Step 3: Follow a system for studying

A. *Preview the material to be learned.*

If you know in advance what part of the textbook is to be covered in your next lecture, flip through the pages before lecture. Glance at the section headings and illustrations. Make notes on what you think the main points will be. Try to guess how the ideas go together. Being right or wrong is not important. The act of previewing the lecture prepares you to learn *during* the lecture, rather than after. This takes about ten minutes, but it can save hours of study time after the lecture.

B. *Take lecture notes.*

What you learn from a lecture depends largely on the quality of notes you take. In general, the best lecture notes are brief summaries that list the main ideas presented. Phrases are used rather than sentences. Ideally, the notes are in outline form, showing major topics and subtopics. The notes are short, but they include all special conditions that are essential to the main ideas. Good lecture notes also anticipate a follow-up in which the comments are expanded. This is done by writing notes on only one half of the page, or on one of the facing pages in a bound notebook. The remaining space is then available for additional comments.

The first sentence of the preceding paragraph noted "the notes *you* take." You can't get useful notes from lectures you don't attend. Be there.

C. *Organize your lecture notes immediately after lecture.*

This is the crucial time. A student who waits 24 hours before reviewing lecture notes forgets almost half of the material presented in lecture. In contrast, a student who reviews lecture notes within a few hours of the lecture retains almost all of what was said, and furthermore, the retention continues for weeks thereafter. If at all possible, schedule the hour after each chemistry lecture for reviewing and organizing your chemistry notes. Nowhere will you find a better bargain in time and learning. And learning is what's it's all about.

You use the open space in your notebook during the review of the lecture. Write in greater detail the items that were condensed to a few words during the lecture. Check your text for anything you didn't quite understand. Summarize the main points of the lecture.

As you organize your lecture notes, pose potential exam questions to yourself. Keep a separate section of your notebook for this purpose. If you have access to old exams, look through them and find the questions that came from the lecture you are reviewing. Even if you don't have old exams, treat each topic in your lecture notes as an answer to an exam question, then write that exam question, just as if you were an instructor looking through your lecture notes and writing an exam. The very act of writing questions will help you to learn the material. Additionally, the exam questions you write now will help you review for the exam later.

D. *Read and take notes from the textbook.*

Your textbook, *Introductory Chemistry: An Active Learning Approach* and *The Active Learning Workbook* contain proven learning aids. Take time now to become acquainted with them. You will then be ready to use both with maximum efficiency.

Preview A textbook assignment should begin with a quick scanning of the material to pick out the highpoints, so you can develop a preliminary idea of its scope and purpose. Your textbook makes the crucial task quite easy. The specific learning tasks you must perform in each assignment are given by the performance goals. The goals appear throughout the text and are listed by section at the end of each chapter. Read them carefully. The goals are the summary highpoints of each chapter.

Because the order of assignments in a chemistry course can vary from instructor to instructor, the textbook has P/Reviews. The P/Reviews are found in the margins and refer to material that has either been covered in a previous assignment or may be covered in a later assignment. They are most important when reviewing a previous assignment. The section number references from the text are given in case you wish to review the topic. We strongly recommend such a review whenever necessary.

Notetaking Now that you have an idea of what your assignment is about, you are ready to learn. *Learn now, not later.* To help you, the text has *Learn It Now!* suggestions printed in the margins in eye-catching blue. Take the hint and do the learning suggested. As you approach each section that has a performance goal, read it carefully and fix in your thought what to look for as you study. The text also has many Target Check questions. These help you to think about concepts and give you instant feedback as to whether or not you understand them. Summarize the main ideas and write them into your notebook in your own words. If what you see with your eye stops over in your mind long enough to be analyzed, revised, and summarized, you are learning it at that time. Continue through the entire assignment this way. When you finish, you will have a compact set of notes covering the main ideas that you have already learned. When test time comes, you will be able to review the main ideas. That is much easier than learning them for the first time the night before the exam.

Unfortunately, many students do not study an assignment in this way. The more common procedure is to sit down with a book and a highlighter pen. Important ideas are marked, not in condensed form, but in their full textbook presentation. Many pages wind up half colored. You don't have to think about something to realize it is important and highlight it. If you don't think about it, you don't learn it. You have only made yourself a promise to learn it later. When test time comes, you have made yourself so many promises that you just can't keep them all. There is just too much to read and too much to learn in too little time.

Do yourself a favor and give your highlighter pens to a child so they can serve their most useful purpose—as a toy. Your notes should be handwritten in your notebook. You can include a reference to textbook pages if you are uncomfortable without having a reference. When you write something down, you are taking the first step toward learning. Remember: If it's important, learn it *now*!

Summaries and Procedures Look for summaries as you study. Summaries appear throughout the textbook, primarily to help you learn, but also to help you in preparing for an exam. These summaries force you to become involved by completing the thoughts that summarize the chapter material. Many chapters also have step-by-step procedures that help you solve problems. Also, don't forget the list of all performance goals and key terms and concepts at the end of each chapter.

Example Problems The only way to learn how to solve chemistry problems is to solve some. The examples in the textbook are distinctively set off from the rest of the text. The beginning of each example states the question. It's just like an exam question; can you solve it? Put your tear-out periodic table shield over the rest of the example, and then go to work at it. When you've got an answer, compare it to the worked-out answer under the shield. If you are self-disciplined and avoid the temptation to look at the solution before you answer the question, you are taking advantage of a very effective learning tool.

As you begin to learn how to solve chemistry problems, it helps to see clearly that your purpose is *not* to solve the problem, but to *learn how to solve* the problem. You are never finished with an assigned problem until you understand it well enough to solve all other problems of that type.

Keep your objective in mind. Your purpose is to learn how to solve problems, not simply to get a correct answer and complete an assignment.

Appendices and Glossary Many students in this course are surprised to learn how much problem solving there is. In addition to Chapter 3 that reviews algebra and problem solving, Appendix I-A helps you with calculator use and Appendix I-B reviews the math skills needed in this course. Learn well the section on chain calculations with calculators. Students who think that they "know" how to use calculators are often awkward (and incorrect) in these operations.

Instructors often use words that are unfamiliar to students, and then expect students to use the same words intelligently. The Glossary is a specialized dictionary that helps to solve the jargon problem. Use both the Glossary and the list of Key Terms and Concepts at the end of each chapter regularly.

E. *Work the end-of-chapter Questions, Exercises, and Problems.*

Notice that the end-of-chapter Questions, Exercises, and Problems (QEP) are fifth in our list of a system for studying chemistry. Does this mean that the QEP are not important? Not at all; they may even be the most important step in your system. But in order to use them most efficiently, you must complete everything that comes before them in order. Avoid the temptation to jump into the QEP without learning from the lecture and textbook first. You must learn the material before you attempt to apply it.

Many instructors will assign QEPs from this workbook for each chapter. In this case, do all of the parallel questions in the textbook first. Check your answers at the end of the chapter. If your answer does not match ours, go back to the section from which the question is based. Study the material again. Diagnose what you didn't learn the first time and learn it NOW. Keep in mind the objective while answering questions: You are learning how to solve problems. Simply arriving at the correct answer is not sufficient. Once you have mastered the concepts behind the textbook questions, do the workbook questions.

Some instructors do not assign QEPs. They expect you to figure out for yourself what questions you should practice with. If this is the case for your course, we suggest you try, as a minimum, the questions for which we have provided answers for in the textbook.

F. *Review.*

The final step in your study system is a review. How often you review will depend on the demands of your course. In most cases, you will have periodic quizzes or exams. These provide you with a series of review intervals. Generally, you will schedule a review study period before every exam. If you have been writing potential exam questions from your lecture notes, as we suggested, when a review session is needed, you have a ready-made sample exam. Finding another student who prepares potential exam questions and exchanging those questions, answering them individually in exam-like conditions (no outside references; limited time similar to that allowed in your course), and analyzing your answers makes an ideal exam review.

Customizing Your Study Plan

Your general study plan needs to be custom designed depending on the type of material being studied. The general techniques we discussed previously work for all types of chemistry content, but you should not study conceptual material in the same way you study the techniques of solving chemistry problems. Here we will discuss studying the two broad categories of chemistry content: concepts and problems.

Chemistry Concepts

A concept is an idea that usually can be summarized with a word or phrase. "Dog" is an example concept that you are familiar with. Not all dogs are the same, but you have a general idea of the characteristics of all dogs. Your mental picture of *dog* is good enough for you to identify an unfamiliar breed of dog as a dog, even if you have never seen that kind of dog before. Now try to define *dog* in words. It's difficult, isn't it? You know what a dog is, but giving a verbal description is difficult. A challenge in academic courses is to translate something that you have a good feel for, like the definition of dog, into a concise and accurate written definition.

When you learn chemistry, the mental challenge is usually reversed. You are given a word or phrase that represents a concept, along with its definition, and then you are expected to recognize examples of that concept. It's easy to memorize and recite the definition, but it's difficult to apply that definition. Most chemistry instructors will expect you to do just that: apply the definition. Do you see the difference between everyday learning and school learning? With dogs, you see many individual dogs over your lifetime and you form a general concept of dog. With chemistry concepts, you will be given a general concept and then be expected to recognize examples—and nonexamples (something we'll discuss soon)—of that concept.

Learning chemistry concepts is accomplished with three specialized steps. You should follow the normal procedures for studying that we've outlined previously, but pay special attention to these steps when you study concepts:

Learn the definition with mental images Make each definition more than words. Try to imagine a physical representation of the concept. We provide many illustrations and photographs in the text to help you with forming mental images. The static figures in the textbook are only the beginning, however. You need to make these images come alive. Imagine them moving about and interacting with one another. This is the first step toward thinking as a chemist.

Let's say the term *dog* was introduced as a chemical concept. Your lecture notes have the following: "Dog (definition): a domesticated fur-covered mammal generally with four legs a tail, varying in size from a few pounds to over a hundred pounds." When you study the textbook, you would see photos of dogs. You need to put together the textbook photographs with the lecture definition to form a mental image of *dog*.

Learn examples *and* nonexamples of the concept Examples of the concept *dog* include Chihuahua (do you have a mental image?), Golden Retriever, and mutt. What do these examples have in common that places them under the *dog* concept? Look for linking features.

Learning nonexamples is just as important as learning examples; sometimes it is even more important. A nonexample is something not part of the concept. *Cat* is a nonexample for *dog*. Sometimes small children will see a cat for the first time and exclaim, "doggie!" They refine their concept of dog by learning that this nonexample is not to be included in their concept. Nonexamples are only important when they can be confused with the examples of the concept. *Worm* is an unimportant nonexample for the dog concept. *Wolf* is a very important nonexample related to the dog concept.

Look for links among concepts Determining relationships among concepts is a key to understanding each of the individual concepts. We use Concept Linking Exercises in the textbook to help you look for these relationships. You can easily think of some relationships among the concepts *dog*, *animal*, *mammal*, and *pet*. Understanding these relationships strengthens the understanding of each of the individual concepts and how it relates to other concepts. We encourage you to complete the Concept Linking Exercise at the end of each chapter.

Chemistry Problems

Chemistry problems are those types of questions usually referred to as "word problems" in a math course. Most chemistry courses usually involve many, many problems, and they are *all* word problems. You need to approach material intended to teach you how to solve problems in a different manner than you approach conceptual material. You will generally find that chemistry problems can be divided into two classifications: algorithms and complex problems.

Algorithms Algorithms are procedures that can be used to solve a number of different questions of the same general type. When you recognize a question as belonging to a particular category, you can recall and apply a specific procedure to that question. For example, you probably already have an algorithm stored in your memory to solve the algebraic exercise "$3x + 5 = 23$." The procedure is (1) isolate the term with the unknown by adding or subtracting ($3x = 18$); and (2) find the unknown by multiplying or dividing ($x = 6$). When you see a similar exercise, such as "$0.1x - 34 = 44.5$," you apply the same procedure even though the exercise itself has different numbers.

One key to successfully solving chemistry problems is to first learn algorithms for common question types and then learn to recognize questions as being one of the learned types. In the textbook, we help you to learn procedures with Procedure summaries. When you come upon one of these, note that it guides you to solving a common chemistry question type. Procedure summaries also have a title. You can use this title to help you categorize problem types.

A common problem in an introductory chemistry course is to look at every question as being unique. With experience in solving problems, you begin to see that problems can be categorized into a smaller number of types. This insight requires that you solve a sufficient number of problems. There is no substitute. As you answer your assigned questions, always look for common problem types among the questions.

Complex Problems Not all questions will fall into a certain category. These are true *problems*. With questions that can be answered with an algorithm, you read the question, realize the question type, and then apply the learned procedure. Sometimes you will read a question and not know what to do. This is a complex problem. Many instructors will put at least one complex problem on an exam.

The first steps in solving a complex problem are those we discuss in the textbook. Start by listing the *GIVEN* quantities and their units and the *WANTED* units. Try to write a *PER/PATH* among the units. Work from both ends. If you cannot construct a path from *GIVEN* to *WANTED*, try working backwards. Work the path from the wanted to wherever it can lead you. Can you work forward to a common point obtained from working backward?

Mentally imagining and/or drawing a representation of the problem is another useful technique to use on complex problems. Translate any macroscopic quantities into their equivalent particulate quantities. Imagine what will happen at the particulate level. Does that help you to "see" the solution?

The one thing that all students who are good at solving complex problems have in common is that they have a good understanding of algorithmic problems. They gain this skill by practice, practice, and more practice. The more you understand the individual pieces of the puzzle, the more likely you will be able to assemble those pieces to solve a complex problem.

Learning Efficiency

No matter how finely tuned your study plan, one fact is true for every student, and every mortal human, for that matter: Time is limited. No study plan can be organized with an unlimited amount of time. You have only a limited number of minutes available to study and learn chemistry. Thus you need to get the most from those limited minutes.

If you have homework that requires three hours of genuine learning, how many hours will you have to study to accomplish that task? It will be more than three hours; for some a little more, for others, a lot more. How much more depends on your learning efficiency.

Learning efficiency is the ratio of minutes of learning to minutes of study time, multiplied by 100. If you learn 48 minutes in an hour of study, your learning efficiency (LE) is

$$\text{LE} = \frac{\text{min of learning}}{\text{min of study}} \times 100 = \frac{48}{60} \times 100 = 80\%$$

The object is to make the numerator as large as possible—maximize learning—while making the denominator as small as possible—minimize the time spent studying. Let's now look at ways to accomplish this crucial task.

Your study area The ideal study area is a quiet place, free from distractions of sight or sound and fully equipped with everything you need, right where you need it. There is nothing to listen to or look at that interrupts concentration, and no time is lost searching for things that are needed. *An ideal study area is the library.* Dorms and homes are generally poor places to study.

Let's see how the efficiency of a typical student is affected by things known to increase study time without increasing learning. We'll assume in this example that three hours of learning are necessary to review for an exam tomorrow. Jennifer sets aside three hours to review for her chemistry exam.

Jennifer doesn't like to study in the library. It's too quiet. She arrives in the student union and then spends 15 minutes looking through her backpack and purse for her notebook, textbook chapter, pencil, paper, and a calculator.

She meant to organize her notes, but she hasn't found time to get to the store and buy a binder. How did Chapter 4 notes get mixed in with Chapter 5? Five more minutes.

It's not very quiet in the union. The noise makes it hard to concentrate. She keeps overhearing people talking about who is dating whom. Joe...she's seen him around...dating Julie? No, she'd never go out with him. 20 minutes.

Kelly stops by for a visit. "Studying chemistry? Gee, that's too difficult for me! Oh you have an exam tomorrow? Well, I'd better leave you alone, but...." The visit lasts for 15 minutes.

Jennifer's cell phone rings. Robert. What's the assignment for tomorrow? Ten minutes.

Jennifer is getting tired. Good thing she's in the union. Coffee is close by. A double mocha espresso ought to do the trick. Wait a minute, where's her wallet? Ten minutes, minimum.

It really is too noisy here. Jen gets out her personal stereo. She can't find a CD in her backpack, so she listens to the radio. During the remainder of her study time, she accepts into her thinking about 40% of the ten minutes of commercials (4 minutes), half of the eight minutes of DJ chatter (4 minutes), and two minutes of news. Also for about 20 minutes, good songs were on. Total cost, 30 minutes.

The lights flash on and off. The union is closing. Where did the time go?

Distraction	Cost in Minutes
Organizing work area	15
Organizing notes	5
Noise	20
Talking to Kelly	15
Cell phone	10
Coffee	10
Personal stereo	30
Total	105

Wow! That's 105 minutes lost to distractions in a 180-minute study period. Only about 75 minutes of learning were salvaged. Jennifer's learning efficiency is:

$$LE = \frac{\text{min of learning}}{\text{min of study}} \times 100 = \frac{75}{180} \times 100 = 42\%$$

Jen has a problem. Her review required 180 learning minutes, but in three hours of "study" she has only been learning for 75 of those minutes. That leaves 105 minutes to go. But the union is closed, she's tired, and she has an early class tomorrow. If she quits now, her review is only 42% done. If she decides to stay up—maybe go to the local 24-hour restaurant—and continue to study at the same efficiency, she'll need another 250 minutes, or more than four more hours, to complete her three-hour review.

Jen gets a D on her chemistry exam the next day. How could that happen considering that she studied regularly and set aside three hours the night before...?

Surely you are not Jennifer. Even so, isn't there a little bit of her in all of us? Can you do something about that little bit of Jennifer in you?

Fatigue and How to Minimize It Even if you remove all the distractions that surround Jennifer and others like her from your study area, you still must overcome fatigue. After long hours at a task, people become mentally and physically tired. You will not be physically tired if you get enough sleep. If your learning efficiency is high, you will have enough time for adequate sleep. High learning efficiency and enough sleep work together to help you earn better grades.

Mental fatigue is another matter. After lengthy work periods at the same or similar tasks, you lose sharpness and enthusiasm. You must work harder and longer for a given amount of learning. You cannot avoid this type of fatigue completely, but you can minimize its effects.

Try these ideas to minimize fatigue:

1. If you have several subjects to study, first tackle the most difficult or least interesting. When fatigue later starts to appear, you'll be doing the most interesting assignment.

2. Take brief breaks. Study for about 50 minutes, and then take 10 minutes off (but *only* 10 minutes!). Get out of your chair and do something physical. Walk up and down a flight of stairs. Repeat hourly.

3. During your breaks, think about how good you'll feel when you get an A in chemistry. Think about how wonderful it will be when you get your degree. Remind yourself of why you are working so hard.

4. Schedule your study periods. As we have already suggested, you should set aside certain regular periods for study. Limit yourself to those periods, and remind yourself that you need to get the task done in that time period. If you are efficient, you can leave yourself more free time.

Some Closing Thoughts

As we said at the beginning of this study guide chapter, learning chemistry is difficult. This term is the time in your life you have decided to take on this challenge. Don't waste your time. If you set aside enough time to learn chemistry, utilize good learning techniques, and maximize your efficiency, you *will* overcome the challenge of learning chemistry. Now let's get started!

Suggested Assignments for Chapter 1

1. Write a daily schedule for this term. On a sheet of paper, make seven vertical columns, one for each day, and draw horizontal columns for each hour of the day. You'll probably want one sheet for the first eight hours, say 7 AM to 3 PM, and another sheet for the last eight hours, 3 PM to 11 PM. Write in all your scheduled classes, work hours, personal activities, etc. Then schedule two hours of study time for each in-class hour. Modify your schedule throughout the term as necessary.

2. Keep a log of your study time. Use a separate sheet of notebook paper to write down the approximate starting and finishing times for each study session. Estimate your efficiency for each session. Total your study hours each week. Are you spending enough time to meet your goals as a student?

3. Write a study plan for each course you are taking at the beginning of the term. Modify this plan as the term progresses. Look for the similarities and differences among your plans.

Chapter 2

Matter and Energy

☑ TARGET CHECK 2.1-WB

(a) Explain from the standpoint of particle behavior why an Alaska resident can store water outside in a strainer in winter, but not in summer. (b) Why can a closed container be half full of a liquid, but never half full of a gas?

☑ TARGET CHECK 2.2-WB

Classify each of the following properties of lithium as physical (P) or chemical (C) by circling the best choice.

a) Melts at 179°C	P	C
b) Has a metallic luster	P	C
c) Reacts quickly with hot water, giving hydrogen gas	P	C
d) Becomes a white powder if exposed to the air	P	C
e) Conducts electricity	P	C
f) Is stable in helium	P	C

Classify the following changes as physical (P) or chemical (C) by circling the best choice.

g) Breaking a dish	P	C
h) Burning coal	P	C
i) Cooking a fish	P	C
j) Digesting an apple	P	C
k) Melting an ice cube	P	C
l) Decaying garbage	P	C

☑ TARGET CHECK 2.3-WB

Carbon monoxide and carbon dioxide are two gases made from carbon and oxygen. Classify each of the following as a pure substance (P) or mixture (M) by circling the best choice (the E and C columns are used to answer Target Check 2.5-WB).

a) Carbon	P	M	E	C
b) Carbon and oxygen	P	M	E	C
c) Oxygen	P	M	E	C
d) Carbon and carbon dioxide	P	M	E	C
e) Carbon dioxide	P	M	E	C
f) Carbon dioxide and carbon monoxide	P	M	E	C
g) Carbon monoxide	P	M	E	C

☑ TARGET CHECK 2.4-WB

Classify the following samples of matter as homogeneous (Hom) or heterogeneous (Het) by circling the best choice:

a) Freshly opened cola	Hom	Het
b) Concrete	Hom	Het
c) Sawdust	Hom	Het
d) The helium in a balloon	Hom	Het
e) Swiss cheese	Hom	Het
f) Poppy seed rye bread	Hom	Het

☑ TARGET CHECK 2.5-WB

Classify each pure substance in Target Check 2.3-WB as an element (E) or compound (C).

☑ TARGET CHECK 2.6-WB

How are the physical properties of compounds related to the physical properties of the elements making up the compounds?

☑ TARGET CHECK 2.7-WB

Six objects are arranged as shown in the sketch below. Two objects are positively charged (+), two have a negative charge (–), and two are neutral (N). Draw a solid line between every pair of objects that are attracted to each other and a dashed line between every pair that repels each other.

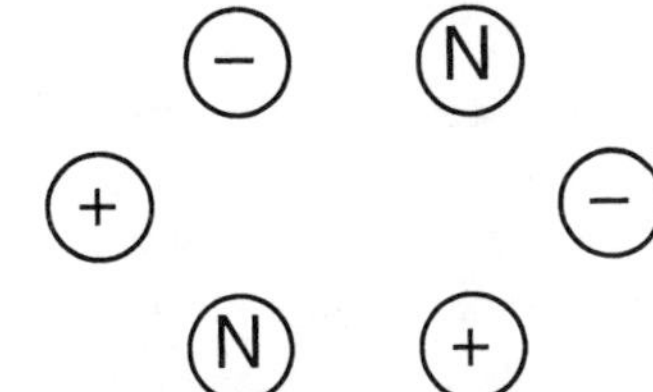

☑ TARGET CHECK 2.8-WB

(i) State whether the following changes are exothermic (Exo) or endothermic (Endo) by circling the best choice:

a) Ice melting	Exo	Endo
b) Food cooking	Exo	Endo
c) Gasoline burning	Exo	Endo
d) Water freezing	Exo	Endo

(ii) A ball is thrown into the air. Compare the relative amounts of potential energy (PE) and kinetic energy (KE) the ball has at the points: (a) as it leaves the thrower's hand, (b) at the very top of its flight, and (c) halfway down.

☑ TARGET CHECK 2.9-WB

If a car uses 1/2 gallon (3.1 lb) of gasoline and 44.7 lb of air to get you to school, what weight of material comes out of the exhaust pipe during the trip?

Name: ________________________ Date: ________________________

ID: ________________________ Section: ________________________

Questions, Exercises, and Problems

Section 2.1: Models and Symbolic Representations of Matter

1. Identify the following samples of matter as macroscopic, microscopic, or particulate: (a) a human skin cell; (b) a sugar molecule; (c) a blade of grass; (d) a helium atom; (e) a single-celled plant too small to be seen with the unaided eye.

2. Suggest a reason for studying matter at the particulate level, given that it is too small to be seen.

Section 2.2: States of Matter: Gases, Liquids, Solids

3. Using spheres to represent individual atoms, sketch particulate illustrations of a substance as it is heated from the solid to the liquid and to the gaseous state.

Name: ______________________ Date: ______________________

ID: ______________________ Section: ______________________

4. The word *pour* is commonly used in reference to liquids, but not to solids or gases. Can a solid or a gas be poured? Why or why not? If either answer is yes, can you given an example?

5. Which of the three states of matter is most easily compressed? Suggest a reason for this.

Section 2.3: Physical and Chemical Properties and Changes

6. Classify each of the following properties as chemical or physical: (a) hardness of a diamond; (b) combustibility of gasoline; (c) corrosive character of an acid; (d) elasticity of a rubber band; (e) taste of chocolate.

7. Which among the following are physical changes? (a) blowing glass; (b) fermenting grapes; (c) forming a snowflake; (d) evaporating dry ice; (e) decomposing a substance by heating it.

Name: ______________________ Date: ______________________

ID: ______________________ Section: ______________________

8. Consider the following properties of potassium selenide, and classify each as either a physical property or a chemical property: (a) it is a solid at room temperature; (b) it is white in color; (c) it turns red when exposed to air; (d) one cubic centimeter of the solid has a mass of 2.9 grams

9. Consider the following properties of acetaldehyde, and classify each as either a physical or a chemical property: (a) it boils at 21°C; (b) one milliliter of the liquid has a mass of 0.78 gram; (c) it burns at a very high temperature; (d) it reacts with hydroxylamine to form acetaldoxime

Section 2.4: Pure Substances and Mixtures

10. Diamonds and graphite are two forms of carbon. Carbon is an element. Chunks of graphite are sprinkled among the diamonds on a jeweler's display tray. Is the material a pure substance or a mixture? Is the display homogeneous or heterogeneous? Justify both answers.

11. Compare the everyday definition of the term *solution*, as it is used to identify a material substance, with the scientific definition of the term. Give three examples of solutions that help to illustrate the contrast.

Name: ______________________ Date: ______________________

ID: ______________________ Section: ______________________

12. Draw a particulate-level sketch of a heterogeneous pure substance.

13. Draw a particulate-level sketch of a homogeneous mixture.

14. Which are the heterogeneous items in the list that follows? (a) sterling silver; (b) freshly opened root beer; (c) popcorn; (d) scrambled eggs; (e) motor oil.

Name: ______________________ Date: ______________________

ID: ______________________ Section: ______________________

15. Some ice cubes are homogeneous and some are heterogeneous. Into which group do ice cubes from your home refrigerator fall? If homogeneous ice cubes are floating on water in a glass, are the contents of the glass homogeneous or heterogeneous? Justify both answers.

16. Apart from food, can you list five things in your home that are homogeneous?

17. Explain why milk from the grocery store is described as "homogenized." What is unhomogenized milk?

Name: ______________________ Date: ______________________

ID: ______________________ Section: ______________________

18. Are the substances in the illustrations in textbook Question 25 homogeneous or heterogeneous?

Section 2.5: Separation of Mixtures

19. Suppose someone emptied ball bearings into a container of salt. Could you separate the ball bearings from the salt? How? Would your method use a physical change or a chemical change?

20. A liquid which may be either pure or a mixture is placed in a distillation apparatus (Figure 2.12). The liquid is allowed to boil and some condenses in the receiving flask. The remaining liquid is then removed and frozen, and the freezing point is found to be lower than the freezing point of the original liquid. Is the original liquid pure or a mixture? Explain.

Name: ______________________ Date: ______________________

ID: ______________________ Section: ______________________

Section 2.6: Elements and Compounds

21. Classify the following as compounds or elements: (a) silver bromide (used in photography); (b) calcium carbonate (limestone); (c) sodium hydroxide (lye); (d) uranium; (e) tin; (f) titanium.

22. Which of the following are elements, and which are compounds? (a) $NaOH$; (b) $BaCl_2$; (c) He; (d) Ag; (e) Fe_2O_3.

23. Can a compound be decomposed into two other compounds?

24. How can you tell if a substance is an element or a compound on the macroscopic level? How can you tell on the particulate level?

Name: ____________________ Date: ____________________

ID: ____________________ Section: ____________________

25. (a) Which of the following substances would you expect to be elements and which would you expect to be compounds? (1) calcium carbonate; (2) arsenic; (3) uranium; (4) potassium chloride; (5) chloromethane.
(b) On what general rule do you base your answers to Part (a)? Can you name any exceptions to this general rule?

26. Metal A dissolves in nitric acid. The original metal can be recovered if Metal B is placed in the aqueous solution. Metal A becomes heavier after prolonged exposure to air. The procedure is faster if the metal is heated. From the evidence given, can you tell if Metal A definitely is or could be an element or a compound? If you cannot, what other information is necessary to make that classification?

27. Consider the following classifications as shown in the table below: gas, liquid, or solid (G, L, S); pure substance or mixture (P, M); homogeneous or heterogeneous (Hom, Het); element or compound (E, C). Place in the table the symbol, such as G, L, or S that best describes each substance in its most common state at room temperature. Assume that the material is clean and uncontaminated. (The first box is filled in as an example.)

	G, L, S	**P, M**	**Hom, Het**	**E, C**
Factory smokestack emissions	All, but mostly G			
Concrete				
Helium in a steel cylinder				
Hummingbird feeder solution				
Table salt				

Name: ______________________ Date: ______________________

ID: ______________________ Section: ______________________

Section 2.7: The Electrical Character of Matter

28. What is the main difference between electrostatic forces and gravitational forces? Which is more similar to the magnetic force? Can two or all three of these forces be exerted between two objects at the same time?

Section 2.8: Characteristics of a Chemical Change

29. Identify the reactants and products in the equation $AgNO_3 + NaCl \rightarrow AgCl + NaNO_3$.

30. In the equation $Ni + Cu(NO_3)_2 \rightarrow Ni(NO_3)_2 + Cu$, which of the reactants is/are elements, and which of the products is/are compounds?

31. Which of the following processes is/are exothermic? (a) water freezing; (b) water vapor in the air changing to liquid water droplets on a windowpane; (c) molten iron solidifying; (d) chocolate candy melting.

Name: ______________________ Date: ______________________

ID: ______________________ Section: ______________________

32. As a child plays on a swing, at what point in her movement is her kinetic energy the greatest? At what point is potential energy at its maximum?

Section 2.9: Conservation Laws and Chemical Change

33. The rock that remains after solid limestone is heated weighs less than the original limestone. What do you conclude has happened?

34. Identify several energy conversions that occur regularly in your home. State whether each is useful, wasteful, or sometimes useful and sometimes wasteful.

Name: ______________________ Date: ______________________

ID: ______________________ Section: ______________________

General Questions

35. A natural-food store advertises that no chemicals are present in any food sold in the store. If their ad is true, what do you expect to find in the store?

36. Name some pure substances you have used today.

37. How many homogeneous substances can you reach without moving from where you are sitting right now?

Name: ______________________ Date: ______________________

ID: ______________________ Section: ______________________

38. Which of the following can be pure substances: mercury, milk, water, a tree, ink, iced tea, ice, carbon?

39. Can you have more than one compound made of the same two elements? If yes, try to give an example.

40. Rainwater comes from the oceans. Is rainwater more pure, less pure, or of the same purity as ocean water? Explain.

Name: ______________________ Date: ______________________

ID: ______________________ Section: ______________________

More Challenging Problems

41. The density of a liquid is determined in the laboratory. The liquid is left in an open container overnight. The next morning the density is measured again and found to be greater than it was the day before. Is the liquid a pure substance or a mixture? Explain your answer.

42. There is always an increase in potential energy when an object is raised higher above the surface of the earth, that is, when the distance between the earth and the object is increased. Increasing the distance between two electrically charged objects, however, may raise or lower potential energy. How can this be?

43. In the gravitational field of the earth, an object always falls until it is stopped by some physical object that prevents it from falling farther. Two electrically charged objects, each of which is made up of unequal numbers of both positive and negative charges, will reach a certain separation distance and stay there without physical support. Can you suggest an explanation for this?

Target Check Answers

1. (a) As solid ice, the particles remain in fixed positions relative to each other and cannot rearrange themselves so they can fit and flow between the openings in the strainer. As a liquid, which the particles would be in the summer temperatures, this flowing is possible. (b) Liquids have definite volume and fill their containers only to the extent of that volume. Gases completely fill any container, regardless of volume.
2. Physical properties (P) are a, b, and e; chemical properties (C) are c, d, and f.
   Physical changes (P) are g and k; chemical changes (C) are h, i, j, and l.
3. Items a, c, e, and g are pure substances (P); items b, d, and f are mixtures (M).
4. The only homogeneous substance is helium, choice d. The others have visibly distinct phases.
5. Items a and c are elements (E); items e and g are compounds (C). Item b is a mixture of two elements; d is a mixture of an element and a compound; f is a mixture of two compounds. The important point is that a compound is a pure substance, one kind of matter, not two. A mixture has two or more kinds of matter, which may be two elements, two compounds, or an element and a compound.
6. There is no relationship between the properties of a compound and the properties of the elements in that compound.
7. Objects with unlike charges attract each other and objects having the same charge repel. Neutral objects experience neither attraction nor repulsion between each other or with charged objects.

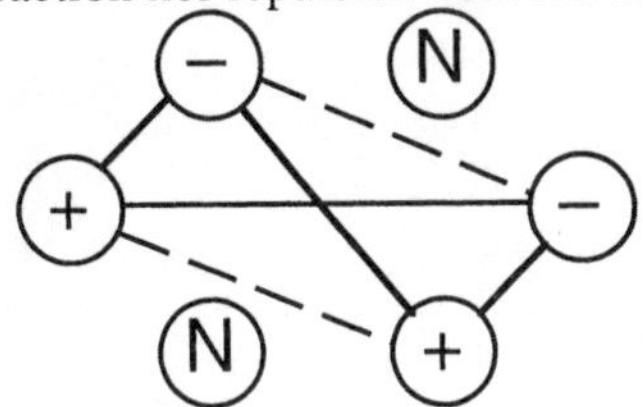

8. (i) The exothermic changes are c and d; the endothermic changes are a and b. You must heat—put energy into—ice to melt it or into food to cook it. There is no question that burning gasoline gives off heat. The release of heat when water freezes is not very apparent, but it is the opposite of ice melting.
   (ii) As it leaves the thrower's hand (a), the ball has low PE, high KE. At the very top of the flight (b), the ball has high PE, low KE. Halfway down, the PE and KE are about equal.
9. The total weight of the exhaust is 47.8 pounds. This illustrates the Law of Conservation of Mass, which states that the mass of reactants in a chemical change equals the mass of the products.

Chapter 3

Measurement and Chemical Calculations

EXAMPLE 3.1-WB

Write each of the following in exponential notation:

(a) 38,981 =

(b) 0.00000339 =

(c) 0.00037 =

(d) 33.087 =

(a) 3.8981×10^4
(b) 3.39×10^{-6}
(c) 3.7×10^{-4}
(d) 3.3087×10^1

EXAMPLE 3.2-WB

Write each of the following numbers in ordinary decimal form:

(a) 2.2×10^5 =

(b) 0.0887×10^{-3} =

(c) 32.9×10^3 =

(d) 3.017×10^{-6} =

(a) 220,000
(b) 0.0000887
(c) 32,900
(d) 0.000003017

EXAMPLE 3.3-WB

Perform each of the following calculations. Our answers are rounded off to three digits, beginning with the first nonzero digit.

$(6.73 \times 10^{-3})(9.11 \times 10^{-3}) =$

$\dfrac{5.08 \times 10^{-3}}{7.23 \times 10^{-9}} =$

$(2.93 \times 10^5)(4.85 \times 10^6)(5.58 \times 10^{-3}) =$

$\dfrac{(3.05 \times 10^{-6})(2.19 \times 10^{-3})}{5.48 \times 10^{-5}} =$

$(6.73 \times 10^{-3})(9.11 \times 10^{-3}) = 6.13 \times 10^{-5}$

$\dfrac{5.08 \times 10^{-3}}{7.23 \times 10^{-9}} = 7.03 \times 10^5$

$(2.93 \times 10^5)(4.85 \times 10^6)(5.58 \times 10^{-3}) = 7.93 \times 10^9$

$\dfrac{(3.05 \times 10^{-6})(2.19 \times 10^{-3})}{5.48 \times 10^{-5}} = 1.22 \times 10^{-4}$

EXAMPLE 3.4-WB

Add or subtract the following numbers:

$4.82 \times 10^{-3} + 9.37 \times 10^{-5} =$

$9.22 \times 10^{6} - 8.38 \times 10^{5} =$

$4.82 \times 10^{-3} + 9.37 \times 10^{-5} = 4.9137 \times 10^{-3}$
$9.22 \times 10^{6} - 8.38 \times 10^{5} = 8.382 \times 10^{6}$

EXAMPLE 3.5-WB

How many years are in 156 weeks?

This problem is similar to Example 3.5 in the textbook, although this answer may not be as immediately apparent. Write the *GIVEN* quantity and units and the *WANTED* units.

GIVEN: 156 weeks *WANTED:* years

Now write the *PER* expression that links weeks and years. In other words, how many weeks are in a year?

---

52 weeks per year

Use the information you've written out so far to write the *PER/PATH* for this problem.

PER/PATH: weeks $\xrightarrow{\text{52 weeks/year}}$ years

You may have written 1 year/52 weeks above the arrow. This is also correct. Our *PER* expressions do not necessarily indicate whether the conversion factor will be exactly as written above the *PER/PATH* arrow or its inverse. We make this decision in the problem setup itself.

You are ready to write the problem setup. Do so, and show how the units cancel algebraically, leaving only the *WANTED* units.

$$156 \text{ weeks} \times \frac{1 \text{ year}}{52 \text{ weeks}} =$$

The weeks cancel, leaving years as the surviving unit. This matches the *WANTED* unit, indicating that the problem setup is correct.

Complete the problem by calculating the answer. Write it and its units in the space above.

$$156 \text{ weeks} \times \frac{1 \text{ year}}{52 \text{ weeks}} = 3 \text{ years}$$

EXAMPLE 3.6-WB

A 5.5-fathom ship channel has been dug through a shallow stretch of the Mississippi River. If a fathom is six feet, how deep is the channel, in inches?

Fathoms are most likely an unfamiliar unit to you, but the problem statement defines it in terms of the equivalent number of feet. Notice, however, that the answer is to be expressed in inches. You know the conversion between feet and inches, and thus you can complete the solution in two steps. Write the *GIVEN, WANTED, AND PER/PATH.*

GIVEN: 5.5 fathoms *WANTED:* inches

PER/PATH: fathoms $\xrightarrow{\text{1 fathom/6 feet}}$ feet $\xrightarrow{\text{12 inches/foot}}$ inches

For most quantitative problems, when you've reached this point in the problem-solving process, you've done the bulk of the work. The steps that follow simply apply the reasoning you done so far. You are ready to setup the problem, cancel the units, and calculate the final answer.

$$5.5 \text{ fathoms} \times \frac{6 \text{ feet}}{\text{fathom}} \times \frac{12 \text{ inches}}{\text{foot}} = 396 \text{ inches}$$

Does your answer make sense? Should there be more inches than fathoms?

More inches (smaller unit) than fathoms (larger unit). The larger/smaller reasoning checks.

EXAMPLE 3.7-WB

How many hours will it take an airplane to travel 1183 miles at an average speed of 455 miles per hour?

Write the *GIVEN, WANTED,* and *PER/PATH*. The speed of the airplane is the *PER* expression for this problem.

GIVEN: 1183 miles *WANTED:* hours

PER/PATH: miles $\xrightarrow{\text{455 miles/hour}}$ hours

Complete the solution by writing the setup and answer. Check your answer.

$$1183 \text{ miles} \times \frac{1 \text{ hour}}{455 \text{ miles}} = 2.6 \text{ hours}$$

Check: More miles than hours. OK.

An airplane should cover more than 1 mile per hour! The answer makes sense.

EXAMPLE 3.8-WB

There are 2 tablespoons in 1 fluid ounce, 8 fluid ounces in one cup, 4 cups in 1 quart, and 4 quarts in 1 gallon. How many tablespoons are there in 2.31 gallons of water?

Don't be intimidated by all of the conversion factors given in this problem. You will take it step by step, and in doing so, you will see how dimensional analysis allows you to solve problems with many unit conversions. Give yourself a clear picture of where you are starting from and when you want to go by writing the *GIVEN* and *WANTED.*

GIVEN: 2.31 gallons *WANTED:* tablespoons

Now you need to construct a *PER/PATH* from gallons to tablespoons. Your strategy will be to find a conversion factor in the problem statement that will allow you to convert from gallons to some other unit. You will then connect that unit to another, and continue until you get to tablespoons. Don't worry if you take a false turn or two as you work this out; problem solving is often messy. Write the *PER/PATH.*

PER/PATH: gallons $\xrightarrow{\text{4 quarts/gallon}}$ quarts $\xrightarrow{\text{4 cups/quart}}$ cups $\xrightarrow{\text{8 fluid ounces/1 cup}}$ fluid ounces $\xrightarrow{\text{2 tablespoons/1 fluid ounce}}$ tablespoons

Completing the problem is accomplished by writing the setup, canceling the units, and calculating the final answer.

$$2.31 \text{ gallons} \times \frac{4 \text{ quarts}}{\text{gallon}} \times \frac{4 \text{ cups}}{\text{quart}} \times \frac{8 \text{ fluid ounces}}{\text{cup}} \times \frac{2 \text{ tablespoons}}{\text{cup}} = 591.36 \text{ tablespoons}$$

EXAMPLE 3.9-WB

How many seconds are needed for a race horse to run six furlongs at 30.7 miles per hour? There are 40 rods in one furlong, 5.5 yards in one rod, 3 feet in one yard, and 5280 feet per mile.

This is another problem with a lot of conversion factors. If you apply the dimensional analysis problem-solving method systematically, you will be able to arrive at the correct solution. Remember, don't expect to immediately see the path from *GIVEN* to *WANTED*. You may have to work with the units a bit. We'll let you go all the way without intermediate assistance this time. Write the complete solution, including the answer.

GIVEN: 6 furlongs *WANTED:* seconds

PER/PATH: furlongs $\xrightarrow{\text{40 rods/furlong}}$ rods $\xrightarrow{\text{5.5 yards/rod}}$ yard $\xrightarrow{\text{3 feet/yard}}$ feet $\xrightarrow{\text{5280 feet/mile}}$ miles $\xrightarrow{\text{30.7 miles/hour}}$ hours $\xrightarrow{\text{60 minutes/hour}}$ minutes $\xrightarrow{\text{60 seconds/minute}}$ seconds

$$6 \text{ furlongs} \times \frac{40 \text{ rods}}{\text{furlong}} \times \frac{5.5 \text{ yards}}{\text{rod}} \times \frac{3 \text{ feet}}{\text{yard}} \times \frac{1 \text{ mile}}{5280 \text{ feet}} \times \frac{1 \text{ hour}}{30.7 \text{ miles}} \times \frac{60 \text{ minutes}}{\text{hour}} \times \frac{60 \text{ seconds}}{\text{minute}} = 88 \text{ seconds}$$

The problem clearly illustrates that dimensional analysis is a powerful problem-solving technique. It will prove to be quite helpful when you are working problems with newly-learned chemistry units.

EXAMPLE 3.10-WB

Calculate the number of meters in 1.64 kilometers.

The purposes of this example are (1) to give you a chance to try for the answer simply by moving the decimal point, (2) to give you practice in solving the problem by dimensional analysis, and (3) to help you decide which method is best for you. First, what do you think the answer is?

(1.64×1000) m, which is 1640 m Check: Larger/smaller comparison is OK.
There are 1000 meters in a kilometer, that is, 1000 times as many meters as the number of kilometers, so $1.64 \times 1000 = 1640$. The decimal moves three places to the right. If you answered 0.00164 m, you will be particularly interested in the discussion after the example.

Now set up and solve the problem by dimensional analysis.

GIVEN: 1.64 km *WANTED:* m *PER/PATH:* km $\xrightarrow{1000\text{ m/km}}$ m

$$1.64\text{ km} \times \frac{1000\text{ m}}{\text{km}} = 1640\text{ m}$$

Check: More meters (smaller unit) than kilometers (larger unit). OK.

EXAMPLE 3.11-WB

Convert 6.65×10^{-5} kg to mg.

Write the *GIVEN, WANTED,* and *PER/PATH.* Convert from the *GIVEN* unit to the base unit, and then from the base unit to the *WANTED* unit.

GIVEN: 6.65×10^{-5} kg *WANTED:* mg

PER/PATH: kg $\xrightarrow{1000\text{ g/kg}}$ g $\xrightarrow{1000\text{ mg/g}}$ mg

Set up and solve the problem.

$$6.65 \times 10^{-5}\text{ kg} \times \frac{1000\text{ g}}{\text{kg}} \times \frac{1000\text{ mg}}{\text{g}} = 66.6\text{ mg}$$

EXAMPLE 3.12-WB

Express 27,935 cm^3 in mL, L, and kL.

Take it all the way without assistance.

27,935 cm^3 = 27,935 mL because 1 cm^3 = 1 mL.

GIVEN: 27,935 mL *WANTED:* L *PER/PATH:* $\text{mL} \xrightarrow{1000\text{ mL/L}} \text{L}$

$$27{,}935\text{ mL} \times \frac{1\text{ L}}{1000\text{ mL}} = 27.935\text{ L}$$

GIVEN: 27.935 L *WANTED:* kL *PER/PATH:* $\text{L} \xrightarrow{1000\text{ L/kL}} \text{kL}$

$$27.935\text{ L} \times \frac{1\text{ kL}}{1000\text{ L}} = 0.027935\text{ kL}$$

EXAMPLE 3.13-WB

How many significant figures are in each of the following quantities?

163 mL _____ 0.60 ft _____ 62,700 cm _____ 5.890×10^5 L _____

---

163 mL __3__ 0.60 ft __2*__ 62,700 cm __?†__ 5.890×10^5 L __4*__

Notes: *The final zeros after the decimal identify them as uncertain digits.

†Exponential notation must be used to show if the 7, the first 0, or the second 0 is the doubtful digit:

6.27	$\times 10^4$ is doubtful in hundreds	(three significant figures)
6.270	$\times 10^4$ is doubtful in tens	(four significant figures)
6.2700	$\times 10^4$ is doubtful in ones	(five significant figures)

EXAMPLE 3.14-WB

Round off each of the following quantities to three significant figures.

a) 0.0074562 kg ____________ c) 3.605×10^{-7} L ____________

b) 2.103×10^{4} mm ____________ d) 3.5000 g ____________

---

a) 0.00746 kg or 7.46×10^{-3} kg c) 3.61×10^{-7} L
b) 2.10×10^{4} mm d) 3.50 g

EXAMPLE 3.15-WB

In an experiment in which oxygen is produced by heating potassium chlorate in the presence of a catalyst, a student assembled and weighed a test tube, test tube holder, and catalyst. She then added potassium chlorate and weighed the assembly again. The data were as follows:

Test tube, test tube holder, catalyst, and potassium chlorate	26.255 g
Test tube, test tube holder, and catalyst	24.05 g

The mass of potassium chlorate is the difference between these numbers. Express this mass in the proper number of significant figures.

	26.25	5 g
–	24.05	g
	2.20	5 g = 2.21 g

The 24.05 is doubtful in the hundredths column, so the difference is rounded off to hundredths.

EXAMPLE 3.16-WB

Assuming the numbers are derived from experimental measurements, solve

$$\frac{(2.86 \times 10^{4})(3.163 \times 10^{-2})}{1.8} =$$

and express the answer in the correct number of significant figures.

5.0×10^{2}

The answer should not be shown as 500, because the number of significant figures could be read as one, two, or three. Two significant figures are set by the 1.8.

EXAMPLE 3.17-WB

What is the mass of a sample, expressed in grams, if a milligram balance display reports the sample mass as 23.0 mg?

Set up and solve the problem. Carefully consider significant figures before you write your final answer.

GIVEN: 23.0 mg *WANTED:* g *PER/PATH:* mg $\xrightarrow{1000\text{ mg/g}}$ g

$$23.0\text{ mg} \times \frac{1\text{ g}}{1000\text{ mg}} = 0.0230\text{ g}$$

The measured quantity in this problem, 23.0 mg, has 3 significant figures. The *PER* expression is a definition, and this it is exact, infinitely significant. The product is therefore a three-significant-figure number. Even though your calculator display shows 0.23 as the answer, you must add the final zero to correctly express a three-significant-figure answer.

EXAMPLE 3.18-WB

An experiment is conducted to measure the change in mass of a certain reactant over time during a chemical reaction. The results are to be reported in grams per minute. If 0.0390 g of the reactant is present at the beginning of the experiment, and 0.00752 g is present after 8.50 minutes, what is the change in mass divided by time for this reactant?

The question asks for the result of the calculation $\frac{\text{final mass} - \text{initial mass}}{\text{time}}$. Find the answer, but be careful with significant figures.

$$\frac{\text{final mass} - \text{initial mass}}{\text{time}} = \frac{0.00752\text{ g} - 0.0390\text{ g}}{8.50\text{ min}} = \frac{-0.0315\text{ g}}{8.50\text{ min}} = -0.00371\text{ g/min}$$

Our final answer was calculated from the fraction immediately preceeding it. If you left the unrounded difference 0.00752 g – 0.0390 g = –0.03148 g in your calculator and then divided by 8.50 min, your answer would be –0.00370 g/min. This is also acceptable. The textbook paragraph immediately following the instruction to work this example discusses rounding final answers.

EXAMPLE 3.19-WB

An American tourist planning a vacation in Canada learns from a website that the distance between Toronto and Montreal is 555 km. How many miles is this?

Find the most convenient conversion factor for this problem in the Length column of textbook Table 3.3, and then complete the example.

GIVEN: 555 km *WANTED:* miles *PER/PATH:* km $\xrightarrow{1\text{ mi}/1.61\text{ km}}$ mi

$$555\text{ km} \times \frac{1\text{ mi}}{1.61\text{ km}} = 345\text{ mi}$$

Check: A mile is longer than a kilometer. There are more kilometers (smaller unit) than miles (larger unit). OK.

EXAMPLE 3.20-WB

While on a visit to Ontario, Canada, you hear on the weather report that the current temperature is 11 degrees. What is the Fahrenheit temperature?

Equation 3.7 is given as $T_{°F} - 32 = 1.8\ T_{°C}$. Solve this for $T_{°F}$.

$T_{°F} = 1.8\ T_{°C} + 32$

Now substitute and solve in the space above.

$T_{°F} = 1.8\ T_{°C} + 32 = 1.8(11) + 32 = 52°F$

EXAMPLE 3.21-WB

Calculate the Celsius and Fahrenheit temperatures that correspond to 298 kelvins.

Determine the Celsius temperature first.

$T_K = T_{°C} + 273$, so $T_{°C} = T_K - 273 = 298 - 273 = 25°C$

Now convert from Celsius to Fahrenheit.

$T_{°F} - 32 = 1.8\ T_{°C} \Rightarrow T_{°F} = 1.8\ T_{°C} + 32 = 1.8(25) + 32 = 77°F$

EXAMPLE 3.22-WB

A rectangular block of lead measures 5.0 cm by 6.2 cm by 2.1 cm. It has a mass of 742 g. Use these data to determine the density of lead.

By definition, density is mass per unit volume. The problem gives you the mass of the lead block, but it doesn't directly give volume. The volume of a rectangular block is length times width times height. What is the volume of the block?

$V = \ell \times w \times h = 5.0\text{ cm} \times 6.2\text{ cm} \times 2.1\text{ cm} = 65\text{ cm}^3$

The units cm × cm × cm give cm^3, the expected volume unit. All measured quantities are given to two significant figures, so their product is also justified to two significant figures.

Apply the definition of density to arrive at the final answer.

$$D \equiv \frac{m}{V} = \frac{742\text{ g}}{65\text{ cm}^3} = 11\text{ g/cm}^3$$

EXAMPLE 3.23-WB

What is the mass of 1.0 L of air?

At first glance, it may seem that there is not enough information in the problem statement to answer the question. But think about the question for a moment and recall that Table 3.4 in the textbook gives densities of many substances, including air. Notice that the question asks for mass and gives volume. The link between mass and volume is density. Look up the density of air in Table 3.4, and then solve the problem.

$$1.0\text{ L} \times \frac{0.0012\text{ g}}{\text{cm}^3} \times \frac{1\text{ cm}^3}{1\text{ mL}} \times \frac{1000\text{ mL}}{\text{L}} = 1.2\text{ g}$$

EXAMPLE 3.24-WB

The SI unit of heat energy is the joule (J). The more familiar heat unit in the United States is the calorie (cal). Calculate the number of calories in 6.56×10^5 J.

The metric–USCS conversion factor table (Table 3.3) gives you additional information. To *PLAN* your solution to this problem, you must decide if you will solve it by algebra or by dimensional analysis. Which do you choose, and why?

Solve the problem by dimensional analysis.
There is no algebraic equation you can use, but there is a *PER* expression available from Table 3.3.

Complete the problem.

GIVEN: 6.56×10^5 J *WANTED*: cal

PER/PATH: $\text{J} \xrightarrow{4.184\ \text{J/cal}} \text{cal}$

$$6.56 \times 10^5\ \text{J} \times \frac{1\ \text{cal}}{4.184\ \text{J}} = 1.57 \times 10^5\ \text{cal}$$

EXAMPLE 3.25-WB

College chemistry departments buy sulfuric acid by the case. A case contains six bottles, each of which holds 2.5 L of acid. If the storeroom manager decides to buy a minimum of 125 L of sulfuric acid, how many cases should she order?

Can you *PLAN* your solution to this problem without leading questions? Carry the *PLAN* as far as you can.

GIVEN: 125 L *WANTED*: cases

You may have gone farther with your *PLAN* than we have in this first step. If you haven't seen how to solve the problem, ask yourself a few questions. Are the *GIVEN* and *WANTED* quantities related by an equation? Can the *GIVEN* and *WANTED* quantities be linked by one or more *PER* expressions and conversion factors? The answers should tell you whether to use algebra or dimensional analysis. Complete the *PLAN* and solve the problem, as necessary, in the space above.

GIVEN: 125 L *WANTED*: cases

PER/PATH: L $\xrightarrow{\text{1 bottle/2.5 L}}$ bottles $\xrightarrow{\text{1 case/6 bottles}}$ cases

$$125\text{ L} \times \frac{1\text{ bottle}}{2.5\text{ L}} \times \frac{1\text{ case}}{6\text{ bottles}} = 8.3\text{ cases} = 9\text{ cases}$$

The 125 L is a minimum quantity, 2.5 L is a two-significant figure number, and 6 is a counting number, infinitely significant. The answer in cases must be a counting number, an integer. To round off to two significant figures (8.3) means buying a fraction of a case. The normal roundoff of 8.3 to 8 would yield 8 × 6 × 2.5 = 120 L, which is short of the 125-L minimum. Therefore the answer is rounded up to the next integer, 9.

EXAMPLE 3.26-WB

In Example 3.26 in the textbook you calculated that you would have to work 6 weeks to earn enough money to buy a $734.26 stereo system. You would be working five shifts of 4 hours each at $7.25/hr. But, alas, when you received your first paycheck, you found that exactly 23% of your earnings had been withheld for social security, federal and state income taxes, and workman's compensation insurance. Taking these into account, how many weeks are needed to earn the $734.26?

Start by calculating your total take-home pay after 23% is deducted.

$$\frac{\$7.25\text{ earned}}{\text{hr}} \times \frac{\$(100 - 23)\text{ take home}}{\$100\text{ earned}} = \$5.5825\text{/hour take-home pay}$$

You may have calculated 23% of the hourly wage and then subtracted it from the wage. This approach is also correct.

Now find how long you need to work.

$$\$734.26 \times \frac{1\text{ hr}}{\$5.5825} \times \frac{1\text{ shift}}{4\text{ hr}} \times \frac{1\text{ week}}{5\text{ shifts}} = 6.58\text{ weeks} = 7\text{ weeks}$$

Name: ______________________ Date: ______________________

ID: ______________________ Section: ______________________

Questions, Exercises, and Problems

Section 3.2: Exponential (Scientific) Notation

1. Write the following numbers in exponential notation:

   a) 0.000322 Answer: [ ]

   b) 6,030,000,000 Answer: [ ]

   c) 0.00000000000619 Answer: [ ]

2. Write the following exponential numbers in the usual decimal form:

   a) 5.12×10^{6} Answer: [ ]

   b) 8.40×10^{-7} Answer: [ ]

   c) 1.92×10^{21} Answer: [ ]

3. Complete the following operations:

   a) $(7.87 \times 10^{4})(9.26 \times 10^{-8}) =$ Answer: [ ]

   b) $(5.67 \times 10^{-6})(9.05 \times 10^{-7}) =$ Answer: [ ]

   c) $(309)(9.64 \times 10^{6}) =$ Answer: [ ]

   d) $(4.07 \times 10^{3})(8.04 \times 10^{-8})(1.23 \times 10^{-2}) =$ Answer: [ ]

4. Complete the following operations:

   a) $\dfrac{6.18 \times 10^{4}}{817} =$ Answer: [ ]

   b) $\dfrac{4.91 \times 10^{6}}{5.22 \times 10^{5}} =$ Answer: [ ]

   c) $\dfrac{4.60 \times 10^{7}}{1.42 \times 10^{3}} =$ Answer: [ ]

   d) $\dfrac{9.32 \times 10^{4}}{6.24 \times 10^{7}} =$ Answer: [ ]

Name: ____________________ Date: ____________________

ID: ____________________ Section: ____________________

5. Complete the following operations:

a) $\dfrac{9.84 \times 10^3}{(6.12 \times 10^3)(4.27 \times 10^7)} =$ Answer: [ ]

b) $\dfrac{(4.36 \times 10^8)(1.82 \times 10^3)}{0.0856\,(4.7 \times 10^6)} =$ Answer: [ ]

6. Complete the following operations:

a) $6.38 \times 10^7 + 4.01 \times 10^8 =$ Answer: [ ]

b) $1.29 \times 10^{-6} - 9.94 \times 10^{-7} =$ Answer: [ ]

Section 3.3: Dimensional Analysis

Our answers to the questions in this section are rounded off according to the rules given in Section 3.5. Your unrounded answers are acceptable at this time.

7. How long will it take to travel the 406 miles between Los Angeles and San Francisco at an average speed of 48 miles per hour?

Answer: [ ]

8. How many minutes does it take a car traveling 88 km/hr to cover 4.3 km?

Answer: [ ]

9. What will be the cost in dollars for nails for a fence 62 feet long if you need 9 nails per foot of fence, there are 36 nails in a pound, and they sell for 69 cents per pound?

Answer: [ ]

Name: ________________________ Date: ________________

ID: ________________________ Section: ________________

10. An American tourist in Mexico was startled to see $1950 on a menu as the price for a meal. However, that dollar sign refers to Mexican pesos, which on that day had an average rate of 218 pesos per American dollar. How much did the tourist pay for the meal in American funds?

Answer: [ ]

11. How many weeks are in a decade?

Answer: [ ]

Section 3.4: Metric Units

12. A woman stands on a scale in an elevator in a tall building. The elevator starts going up, rises rapidly at constant speed for half a minute, and then slows to a stop. Compare the woman's weight as recorded by the scale and her mass while the elevator is standing still during the starting period, during the constant rate period, and during the slowing period.

13. What is the metric unit of length?

14. Kilobuck is a slang expression for a sum of money. How many dollars are in a kilobuck? How about a megabuck (See Table 3.2)?

Name: ______________________ Date: ______________________

ID: ______________________ Section: ______________________

15. One milliliter is equal to how many liters?

16. Which unit, megagrams or grams, would be more suitable for expressing the mass of an automobile? Why?

Questions 17–19: Make each conversion indicated. Use exponential notation to avoid long integers or decimal fractions. Write your answers without looking at a conversion table.

17. 5.74 cg = ______________ g

 1.41 kg = ______________ g

 4.54×10^{8} cg = ______________ mg

18. 21.7 m = ______________ cm

 517 m = ______________ km

 0.666 km = ______________ cm

19. 494 cm^3 = ______________ mL

 1.91 L = ______________ mL

 874 cm^3 = ______________ L

20. Refer to Table 3.2 for less-common prefixes in these metric conversions.

 7.11 hg = ______________ g

 5.27×10^{-7} m = ______________ pm

 3.63×10^{6} g = ______________ dag

Name: ______________________ Date: ______________________

ID: ______________________ Section: ______________________

Section 3.5: Significant Figures

21. To how many significant figures is each quantity expressed?

75.9 g sugar ____________

89.583 mL weed killer ____________

0.366 in. diameter glass fiber ____________

48,000 cm wire ____________

0.80 ft spaghetti ____________

0.625 kg silver ____________

9.6941×10^{6} cm thread ____________

8.010×10^{-3} L acid ____________

22. Round off each quantity to three significant figures.

6.398×10^{-3} km rope ____________________

0.0178 g silver nitrate ____________________

79,000 m cable ____________________

42,150 tons fertilizer ____________________

$649.85 ____________________

23. A moving-van crew picks up the following items: a couch that weighs 147 pounds, a chair that weighs 67.7 pounds, a piano at 3.6×10^{2} pounds, and several boxes having a total weight of 135.43 pounds. Calculate and express in the correct number of significant figures the total weight of the load.

Answer: [ ]

Name: ____________________ Date: ____________________

ID: ____________________ Section: ____________________

24. A buret contains 22.93 mL sodium hydroxide solution. A few minutes later, however, the volume is down to 19.4 mL because of a small leak. How many milliliters of solution have drained from the buret?

Answer: [ ]

25. The mole is the SI unit for the amount of a substance. The mass of one mole of pure table sugar is 342.3 grams. How many grams of sugar are in exactly $\frac{1}{2}$ mole? What is the mass of 0.764 mole?

Answer: [ ]

26. An empty beaker with a mass of 42.3 g is filled with a liquid, and the resulting mass of the liquid and the beaker is 62.87 g. The volume of this liquid is 19 mL. What is the density of the liquid?

Answer: [ ]

Section 3.6: Metric–USCS Conversions

Questions 27–35: You may consult Table 3.3 while answering these questions.

27. 0.0715 gal = ____________________ cm^3

2.27×10^4 mL = ____________________ gal

Name: ______________________ Date: ______________________

ID: ______________________ Section: ______________________

28. A popular breakfast cereal comes in a box containing 510 g. How many pounds (lb) of cereal is this?

Answer: [ ]

29. The payload of a small pick-up truck is 1450 pounds. What is this in kilograms?

Answer: [ ]

30. There is 115 mg of calcium in a 100 g serving of whole milk. How many grams of calcium is this? How many pounds?

Answer: [ ] [ ]

31. An Austrian boxer reads 69.1 kg when he steps on a balance (scale) in his gymnasium. Should he be classified as a welterweight (136 to 147 lb) or a middleweight (148 to 160 lb)?

Answer: [ ]

32. The height of Angel Falls in Venezuela is 399.9 m. How high is this in (a) yards; (b) feet?

Answer: [ ] [ ]

Name: ______________________ Date: ______________________

ID: ______________________ Section: ______________________

33. The Sears Tower in Chicago is 1454 feet tall. How high is this in meters?

Answer: [ ]

34. The summit of Mount Everest is 29,002 ft above sea level. Express this height in kilometers.

Answer: [ ]

35. An office building is heated by oil-fired burners that draw fuel from a 619 gal storage tank. Calculate the tank volume in liters.

Answer: [ ]

Section 3.7: Temperature

36. Fill in the spaces in the tables below so that each temperature is expressed in all three scales. Round off answers to the nearest degree.

Celsius	Fahrenheit	Kelvin
69		
	–29	
		111
	36	
		358
–141		

Name: ______________________ Date: ______________________

ID: ______________________ Section: ______________________

37. "Normal" body temperature is 98.6°F. What is this temperature in Celsius degrees?

Answer: [ ]

38. Energy conservationists suggest that air conditioners should be set so that they do not turn on until the temperature tops 78°F. What is the Celsius equivalent of this temperature?

Answer: [ ]

39. The world's highest shade temperature was recorded in Libya at 58.0°C. What is its Fahrenheit equivalent?

Answer: [ ]

Section 3.8: Proportionality and Density

40. The amount of heat (Q) absorbed when a pure substance melts is proportional to the mass of the sample (m). Express this proportionality in mathematical form. Change it into an equation, using the symbol ΔH_{fus} for the proportionality constant. This constant is the heat of fusion of a pure substance. If heat is measured in calories, what are the units of heat of fusion? Write a word definition of heat of fusion.

Name: ______________________ Date: ______________________

ID: ______________________ Section: ______________________

41. It takes 7.39 kilocalories to melt 92 grams of ice. Calculate the heat of fusion of water. (See Question 40. Careful on the units; the answer will be in calories per gram.)

Answer: [ ]

42. If the temperature and amount of a gas are held constant, the pressure (P) it exerts is inversely proportional to volume (V). This means that pressure is directly proportional to the inverse of volume, or 1/V. Write this as a proportionality, and then as an equation with k' as the proportionality constant. What are the units of k' if pressure is in atmospheres and volume is in liters?

43. Calculate the density of benzene, an important liquid used in chemistry laboratories, if 166 g of benzene fills a graduated cylinder to the 188-mL mark.

Answer: [ ]

44. Densities of gases are usually measured in grams per liter (g/L). Calculate the density of air if the mass of 15.7 L is 18.6 g.

Answer: [ ]

Name: ______________________ Date: ______________________

ID: ______________________ Section: ______________________

45. Ether, a well-known anesthetic, has a density of 0.736 g/cm^3. What is the volume of 471 g of ether?

Answer: [ ]

46. Determine the mass of 2.0 liters of rubbing alcohol, which has a density of 0.786 g/mL.

Answer: [ ]

General Questions

47. What do you weigh in (a) milligrams; (b) grams; (c) kilograms? Which of these units do you think is best for expressing a person's weight? Why?

48. The density of aluminum is 2.7 g/cm^3. An ecology-minded student has gathered 235 empty aluminum cans for recycling. If there are 19 cans per pound, what will be their volume in cubic centimeters when melted?

Answer: [ ]

Name: ______________________ Date: ______________________

ID: ______________________ Section: ______________________

More Challenging Problems

49. A student's driver's license lists her height as 5 feet, 5 inches. What is her height in meters?

Answer: [ ]

50. The fuel tank in an automobile has a capacity of 11.8 gal. If the density of gasoline is 42.0 lb/ft^3, what is the mass of fuel in kilograms when the tank is full?

Answer: [ ]

51. At high noon on the Lunar equator the temperature may reach 243°F. At night the temperature may sink to –261°F. Express the difference in temperature in degrees Celsius.

Answer: [ ]

52. Calculate the mass in pounds of one gallon of water, given that the density of water is 1.0 g/mL.

Answer: [ ]

Answers to Target Checks

1. 9.43×10^{-4}
2. 399,300
3. 67.0 . . .
4. (a) 4.9137×10^{-3} (b) 8.382×10^{6}
5. a, d, and e are given quantities; b, c, and f are *PER* expressions.
6. 27,935 cm^3 = 27,935 mL = 27.935 L = 0.027935 kL
7. (0.00752 g – 0.0390 g) ÷ 8.50 min = –0.0315 g/8.50 min = –0.00371 g/min
   (See the textbook paragraph just after this Target Check if your answer is slightly different.) By definition, a change in quantity is always calculated by subtracting the initial value from the final value. This is explained in more detail in Section 15.10.
8. $13.5 \text{ gal} \times \frac{3.785 \text{ L}}{\text{gal}} = 51.1 \text{ L}$
9. $1.0 \text{ L air} \times \frac{1000 \text{ mL air}}{\text{L air}} \times \frac{1 \text{ cm}^3 \text{ air}}{1 \text{ mL air}} \times \frac{0.0012 \text{ g air}}{1 \text{ cm}^3 \text{ air}} = 1.2 \text{ g air}$

Chapter 4

Introduction to Gases

EXAMPLE 4.1-WB

Express 746 torr in atmospheres and kilopascals.

Consider the torr-to-atmospheres conversion first. Write the *PLAN* for the solution.

GIVEN: 746 torr *WANTED:* atm *PER/PATH:* torr $\xrightarrow{760\text{ torr/atm}}$ atm

Now complete the calculation.

$$746\text{ torr} \times \frac{1\text{ atm}}{760\text{ torr}} = 0.982\text{ atm}$$

The conversion to kilopascals will finish the example. Equation 4.3 states that 1 atm = 101.3 kPa, so, at first glance, it appears that conversion from the just-calculated 0.982 atm to kPa will be the most straightforward approach. However, note that Equation 4.4 shows that 1 atm = 760 torr. Since 101.3 kPa and 760 torr are both equal to 1 atm, they are equal to one another: 101.3 kPa = 760 torr. This *PER* expression can be used to convert from the 746 torr given in the problem statement to kPa. Either approach is valid. We'll show both solutions. Complete the *PLAN* and calculation.

GIVEN: 0.982 atm *WANTED:* kPa *PER/PATH:* atm $\xrightarrow{101.3\text{ kPa/atm}}$ kPa

$$0.982\text{ atm} \times \frac{101.3\text{ kPa}}{\text{atm}} = 99.5\text{ kPa}$$

GIVEN: 746 torr *WANTED:* kPa *PER/PATH:* torr $\xrightarrow{101.3\text{ kPa/760 torr}}$ kPa

$$746\text{ torr} \times \frac{101.3\text{ kPa}}{760\text{ torr}} = 99.4\text{ kPa}$$

The difference in the answer when comparing the two approaches is acceptable because it reflects differences in rounding in the uncertain digit.

☑ TARGET CHECK 4.1-WB

A fixed quantity of a gas at constant pressure was cooled from 22°C to 0°C. The final volume of the gas is what fraction of the original volume?

EXAMPLE 4.2-WB

A gas is contained in a laboratory cylinder in which the gas volume changes according to the surrounding temperature and pressure. The gas volume was 759 mL at the end of a workday, when the laboratory temperature was 21°C. Overnight, when the heat was turned down, the volume dropped to 744 mL. Assuming that the pressure remained constant, what temperature did the gas reach during the night?

Complete the following table to organize the data in the problem statement.

	Volume	Temperature	Pressure
Initial value (1)			
Final value (2)			

	Volume	Temperature	Pressure
Initial value (1)	759 mL	21°C; 294 K	Constant
Final value (2)	744 mL	T_2	Constant

Write the equation that expresses the Volume–Temperature (Charles's) Law, and then solve it for the unknown variable, T_2.

$$\frac{V_1}{T_1} = \frac{V_2}{T_2} \qquad T_2 = T_1 \times \frac{V_2}{V_1}$$

Finish the problem by solving for the value of T_2. Change the temperature in kelvins to degrees Celsius.

$$T_2 = T_1 \times \frac{V_2}{V_1} = 294\ \text{K} \times \frac{744\ \text{mL}}{759\ \text{mL}} = 288\ \text{K} = 15°\text{C}$$

Does your answer make sense? Does our reasoning approach verify your numerical answer?

$V \propto T$, so if volume decreases, temperature must also decrease. The volume fraction is less than one, which means that the final temperature is less than the initial temperature. The answer is reasonable.

☑ TARGET CHECK 4.2-WB

A fixed quantity of a gas at constant temperature has its volume adjusted so that the pressure decreases from 2.5 atm to 1.0 atm. The final volume of the gas is what multiple of the original volume?

EXAMPLE 4.3-WB

What pressure is required to reduce the volume of a gas from 355 mL to 297 mL if the initial pressure is 685 torr? Temperature and quantity are constant.

Start with a table of initial and final values for volume, temperature, and pressure.

	Volume	Temperature	Pressure
Initial value (1)	355 mL	Constant	685 torr
Final value (2)	297 mL	Constant	P_2

Now write the mathematical statement of the Volume-Pressure (Charles's) Law and rearrange the equation to solve for your unknown, P_2.

$$P_1V_1 = P_2V_2 \qquad P_2 = P_1 \times \frac{V_1}{V_2}$$

Solve for the final pressure.

$$P_2 = P_1 \times \frac{V_1}{V_2} = 685 \text{ torr} \times \frac{355 \text{ mL}}{297 \text{ mL}} = 819 \text{ torr}$$

Finally, check the answer. Is it reasonable? Explain.

$P \propto \frac{1}{V}$, so if one value increases, the other decreases. The volume decreases in the problem, so the pressure must increase. Therefore, the volume fraction must have a value greater than 1. It does, and the pressure increases, so the answer is reasonable.

EXAMPLE 4.4-WB

The gas in a balloon exerts 2.16 atm of pressure and has a volume of 3.19 liters when sitting near a window in a room where the temperature is 18°C. Sunlight comes through the window and heats the gas in the balloon. It expands to 3.42 liters just before it bursts when the pressure reaches 2.26 atm. What was the temperature in the balloon at the time it burst?

Begin with the table:

	Volume	Temperature	Pressure
Initial value (1)			
Final value (2)			

	Volume	Temperature	Pressure
Initial value (1)	3.19 L	18°C; 291 K	2.16 atm
Final value (2)	3.42 L	T_2	2.26 atm

This time the Combined Gas Law equation must be solved for T_2. Do so, and then calculate the temperature in kelvins.

$$\frac{P_1V_1}{T_1} = \frac{P_2V_2}{T_2} \qquad T_2 = T_1 \times \frac{P_2}{P_1} \times \frac{V_2}{V_1} = 291\text{ K} \times \frac{2.26\text{ atm}}{2.16\text{ atm}} \times \frac{3.42\text{ L}}{3.19\text{ L}} = 326\text{ K}$$

Temperature is more meaningful if expressed in Celsius degrees or, to Americans, in Fahrenheit degrees. What are those numbers? (*Hint:* From Section 3.7, the relationship between $T_{°C}$ and $T_{°F}$ is $T_{°F} - 32 = 1.8\ T_{°C}$.)

GIVEN: 326 K *WANTED:* $T_{°C}$

EQUATION: $T_{°C} = T_K - 273 = 326 - 273 = 53°C$

GIVEN: 53°C *WANTED:* $T_{°F}$

EQUATION: $T_{°F} - 32 = 1.8\ T_{°C} = 1.8 \times 53 + 32 = 127°F$

EXAMPLE 4.5-WB

A sample of helium has a volume of 16.3 liters at STP. What volume would it occupy if allowed to reach typical room conditions of 24°C and 0.941 atm?

Set up a table of initial and final values for volume, temperature, and pressure, solve the Combined Gas Law equation for the unknown variable, and complete the calculation.

	Volume	Temperature	Pressure
Initial value (1)	16.3 L	0°C; 273 K	1 atm
Final value (2)	V_2	24°C; 297 K	0.941 atm

$$\frac{P_1V_1}{T_1} = \frac{P_2V_2}{T_2} \qquad V_2 = V_1 \times \frac{P_1}{P_2} \times \frac{T_2}{T_1} = 16.3\text{ L} \times \frac{1.00\text{ atm}}{0.941\text{ atm}} \times \frac{297\text{ K}}{273\text{ K}} = 18.8\text{ L}$$

Check each fraction to see if the answer is reasonable. Does the pressure fraction make sense? How about the temperature fraction?

The pressure decreases, which means that volume will increase and the pressure fraction should be greater than 1. This checks. The temperature increases, and volume and temperature move in the same direction. The temperature fraction must be greater than 1, and it is. The solution is reasonable.

Name: ______________________ Date: ______________________

ID: ______________________ Section: ______________________

Questions, Exercises, and Problems

Section 4.2: The Kinetic Theory of Gases and the Ideal Gas Model

1. What properties of gases are the result of the kinetic character of a gas? Explain.

2. State how and explain why the pressure exerted by a gas is different from the pressure exerted by a liquid or solid.

3. What are the desirable properties of air in an air mattress used by a camper? Show which part of the ideal gas model is related to each property.

Questions 4 through 7: Explain how the physical phenomenon described is related to one or more of the features of the ideal gas model.

4. Pressure is exerted on the top of a tank holding a gas, as well as on its sides and bottom.

5. Balloons expand in all directions when blown up, not just at the bottom as when filled with water.

Name: ______________________ Date: ______________________

ID: ______________________ Section: ______________________

6. Even though an automobile tire is "filled" with air, more air can always be added without increasing the volume of the tire significantly.

7. Gas bubbles always rise through a liquid and become larger as they move upward.

Section 4.3: Gas Measurements

8. What does pressure measure? What does temperature measure?

9. Explain how a manometer works.

Name: ______________________ Date: ______________________

ID: ______________________ Section: ______________________

10. Complete the table by converting the given pressure to each of the other pressures.

atm					
psi					
in. Hg					
cm Hg					
mm Hg	785				
torr		124			
Pa			1.18×10^5		
kPa				91.4	
bar					0.977

11. Find the pressure of the gas in the mercury manometer shown if atmospheric pressure is 747 torr.

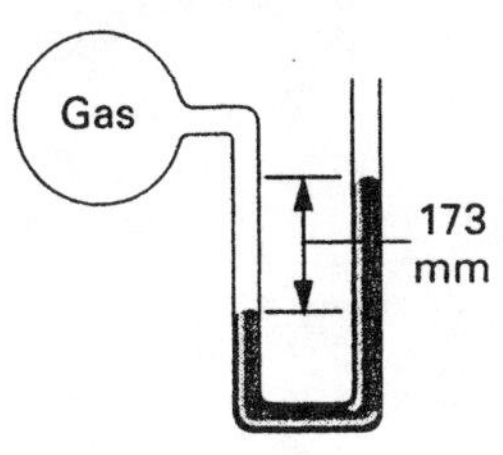

Answer: ______________________

12. A student records a temperature of –18 K in an experiment. What is the nature of things at that temperature? What would you guess the student meant to record? What is the absolute temperature that corresponds to the temperature the student meant to record?

Name: ______________________ Date: ______________________

ID: ______________________ Section: ______________________

13. If the temperature in the room is 31°C, what is the equivalent absolute temperature?

Answer: [ ]

14. Hydrogen remains a gas at very low temperatures. It does not condense to a liquid until the temperature is –253°C, and it freezes shortly thereafter at –259°C. What are the equivalent absolute temperatures?

Answer: [ ]

15. Water in all of its states—solid, liquid, and gas—is present in the world of humans at atmospheric pressure, although at 100°C and higher, humans cannot be exposed to it as a gas. In terms of temperature, humans can be in the presence of hydrogen cyanide in all three states, since it melts at 259 K and changes to a gas at 299 K. It is recommended that you avoid *being* in the presence of hydrogen cyanide, however; it's the deadly gas used in execution chambers. What are the Celsius temperatures at which hydrogen cyanide changes state?

Answer: [ ]

Section 4.4: The Volume–Temperature (Charles's) Law

16. A variable-volume container holds 24.3 L of gas at 55°C. If pressure remains constant, what will the volume be if the temperature falls to 17°C?

Answer: [ ]

Name: ______________________ Date: ______________________

ID: ______________________ Section: ______________________

17. A spring-loaded closure maintains constant pressure on a gas system that holds a fixed quantity of gas, but a bellows allows the volume to adjust for temperature changes. From a starting point of 1.26 L at 19°C, what Celsius temperature will cause the volume to change to 1.34 L?

Answer: [ ]

Section 4.5: The Volume–Pressure (Boyle's) Law

18. If you squeeze the bulb of a dropping pipet (eye dropper) when the tip is below the surface of a liquid, bubbles appear. When the bulb is released, liquid flows into the pipet. Explain why in terms of the Volume–Pressure Law.

19. The pressure on 648 mL of a gas is changed from 772 torr to 695 torr. What is the volume at the new pressure?

Answer: [ ]

20. A cylindrical gas chamber has a piston at one end that can be used to compress or expand the gas. If the gas is initially at 1.22 atm when the volume is 7.26 L, what will the pressure be if the volume is adjusted to 3.60 L?

Answer: [ ]

Name: ______________________ Date: ______________________

ID: ______________________ Section: ______________________

Section 4.6: The Combined Gas Law

21. The gas in a 0.717-L cylinder of a diesel engine exerts a pressure of 744 torr at 27°C. The piston suddenly compresses the gas to 48.6 atm and the temperature rises to 547°C. What is the final volume of the gas?

Answer: [ ]

22. A collapsible balloon for carrying meteorological testing instruments aloft is partly filled with 626 liters of helium, measured at 25°C and 756 torr. Assuming the volume of the balloon is free to expand or contract according to changes in pressure and temperature, what will be its volume at an altitude where the temperature is –58°C and the pressure is 0.641 atm?

Answer: [ ]

23. Why have the arbitrary conditions of STP been established? Are they realistic?

24. If one cubic foot—28.4 L—of air at common room conditions of 23°C and 739 torr is adjusted to STP, what does the volume become?

Answer: [ ]

Name: ______________________ Date: ______________________

ID: ______________________ Section: ______________________

25. An experiment is designed to yield 44.5 milliliters of oxygen, measured at STP. If the actual temperature is 28°C and the actual pressure is 0.894 atm, what volume of oxygen will result?

Answer: [ ]

26. What pressure (atm) will be exerted by a tank of natural gas used for home heating if its volume is 19.6 L at STP and it is compressed to 6.85 L at 24°C?

Answer: [ ]

27. A container with a volume of 56.2 L holds helium at STP. The gas is compressed to 23.7 L and 2.09 atm. To what must the temperature change to satisfy the new volume and pressure?

Answer: [ ]

More Challenging Problems

28. Explain how the following physical phenomenon is related to one or more of the features of the ideal gas model: Very small dust particles, seen in a beam of light passing through a darkened room, appear to be moving about erratically.

Name: ____________________ Date: ________________

ID: ____________________ Section: ________________

29. The volume of the air chamber of a bicycle pump is 0.26 L. The volume of a bicycle tire, including the hose between the pump and the tire, is 1.80 L. If both the tire and the air in the pump chamber begin at 743 torr, what will be the pressure in the tire after a single stroke to the pump?

Answer:

30. A hydrogen cylinder holds gas at 3.67 atm in a laboratory where the temperature is 25°C. To what will the pressure change when the cylinder is placed in a storeroom where the temperature drops to 7°C?

Answer:

31. A gas storage tank is designed to hold a fixed volume and quantity of gas at 1.74 atm and 27°C. To prevent excessive pressure due to overheating, the tank is fitted with a relief valve that opens at 2.00 atm. To what temperature must the gas rise in order to open the valve?

Answer:

32. If 1.62 m^3 of air at 12°C and 738 torr is compressed into a 0.140-m^3 tank, and the temperature is raised to 28°C while the external pressure is 14.7 psi, what pressure (psi) will be read on the gauge?

Answer:

Answers to Target Checks

1. $\frac{V_1}{T_1} = \frac{V_2}{T_2}$ $V_1T_2 = V_2T_1$ $V_2 = V_1 \times \frac{T_2}{T_1}$ $V_{final} = \frac{T_2}{T_1} \times V_{initial}$

$$\frac{T_2}{T_1} = \frac{(0 + 273)\text{ K}}{(22 + 273)\text{ K}} = 0.925$$

2. $P_1V_1 = P_2V_2$ $V_2 = V_1 \times \frac{P_1}{P_2}$ $V_{final} = \frac{P_1}{P_2} \times V_{initial}$

$$\frac{P_1}{P_2} = \frac{2.5\text{ atm}}{1.0\text{ atm}} = 2.5$$

Chapter 5

Atomic Theory: The Nuclear Model of the Atom

☑ TARGET CHECK 5.1-WB

How does Dalton's Atomic Theory account for chemical reactions? How does it support the Law of Conservation of Mass?

☑ TARGET CHECK 5.2-WB

Which subatomic particle, proton, neutron, or electron, is the lightest? Which is the heaviest? Which are charged? Which are not charged?

☑ TARGET CHECK 5.3-WB

The gold foil used in the Rutherford scattering experiment was about 2000 atoms thick. Why did most of the alpha particles pass straight through the solid foil?

EXAMPLE 5.1-WB

The element boron has two naturally-occurring isotopes, one of which has a mass number of 10, and the other has mass number 11. For each isotope, indicate the (a) atomic number, (b) number of protons, (c) number of neutrons, and (d) nuclear charge.

Begin with the atomic number of boron. Look on the reference sheet for boron, and write its atomic number. Also write the symbol for atomic number.

Atomic number = Z = 5

Next, explain the relationship between the atomic number of an element and the number of protons in the nucleus of an atom of that element. Then state the number of protons in atoms of each isotope.

The atomic number of an element *is* the number of protons in the nucleus of an atom of the element. Both isotopes have 5 protons in each atom.

Calculate the number of neutrons in each isotope.

---

11 – 5 = 6 neutrons in boron-11; 10 – 5 = 5 neutrons in boron-10

The phrase *nuclear charge* literally refers to the charge on the nucleus. What is the charge on a nucleus with 5 protons?

5+

☑ TARGET CHECK 5.4-WB

Can isotopes of the same element have the same mass number? the same atomic number? Can isotopes of different elements have the same mass number? the same atomic number? Explain.

EXAMPLE 5.3-WB

Use the data in Table 5.2 to calculate the atomic mass of sulfur (symbol S).

0.950	×	31.97207 amu	=	30.3734665	amu	=	30.4	amu
0.0076	×	32.97146 amu	=	0.250583096	amu	=	0.25	amu
0.0422	×	33.96786 amu	=	1.433443692	amu	=	1.43	amu
0.00014	×	35.96709 amu	=	0.0050353926	amu	=	0.0050	amu
							32.0850	amu = 32.1 amu

EXAMPLE 5.4-WB

List the group (__________), period (__________), symbol (__________), and atomic mass (__________) for lead Z = 82.

Group 4A/14; Period 6; symbol, Pb; atomic mass, 207.2 amu

The first element in Group 4A/14 is carbon; Z = 6, which is in the second period. Counting down from there, Z = 82 is in Period 6.

☑ TARGET CHECK 5.5-WB

(i) Select the correct placement in the periodic table for nitrogen, N.

a) Group 5A/15, Period 1 b) Group 2, Period 5A/15

c) Group 7, Period 2A/2 d) Group 5A/15, Period 2

Answer:

(ii) Select the correct statement about an isotope of an element for which the atomic number is 16 and mass number is 32.

a) The element is sulfur, S, for which Z = 32, Group 6A/16, Period 3

b) The element is sulfur, S, for which A = 32, Group 6A/16, Period 3

c) The element is oxygen, O, for which Z = 16, Group 6A/16, Period 2

d) The element is oxygen, O, for which A = 16, Group 6A/16, Period 2

Answer:

EXAMPLE 5.5-WB

For each element below, write the name; for each name, write the symbol.

P (________________) potassium (________)

Cl (________________) zinc (________)

Fe (________________) bromine (________)

P, phosphorus potassium, K

Cl, chlorine zinc, Zn

Fe, iron bromine, Br

Name: ______________________ Date: ______________________

ID: ______________________ Section: ______________________

Questions, Exercises, and Problems

Section 5.1: Dalton's Atomic Theory

1. According to Dalton's atomic theory, can more than one compound be made from atoms of the same two elements?

2. Show that the Law of Definite Composition is explained by Dalton's atomic theory.

3. The chemical name for limestone, a compound of calcium, carbon, and oxygen, is calcium carbonate. When heated, limestone decomposes into solid calcium oxide and gaseous carbon dioxide. From the names of the products, tell where the atoms of each element may be found after the reaction. How does the atomic theory explain this?

4. Sulfur and fluorine form at least two compounds—SF_4 and SF_6. Explain how these compounds can be used as an example of the Law of Multiple Proportions.

Section 5.2: Subatomic Particles

5. Compare the three major parts of an atom in charge and mass.

Name: ______________________ Date: ______________________

ID: ______________________ Section: ______________________

Section 5.3: The Nuclear Atom

6. How can we account for the fact that, in the Rutherford scattering experiment, some of the alpha particles were deflected from their paths through the gold foil, and some were even bounced back at various angles?

7. What name is given to the central part of an atom?

8. Describe the activity of electrons according to the planetary model of the atom that appeared after the Rutherford scattering experiment.

Section 5.4: Isotopes

9. Can two different elements have the same atomic number? Explain.

10. Explain why isotopes of different elements can have the same mass number, but isotopes of the same element cannot.

Name: ______________________ Date: ______________________

ID: ______________________ Section: ______________________

11. From the information given in the following table, complete as many blanks as you can without looking at any reference. If there are unfilled spaces, continue by referring to your periodic table. As a last resort, check the table of elements on the reference page.

Name of Element	Nuclear Symbol	Atomic Number	Mass Number	Protons	Neutrons	Electrons
					24	21
	$^{76}_{32}Ge$					
			122		72	
			37			17
		11			12	

Section 5.5: Atomic Mass

Although this set of questions is based on material in Section 5.5, some parts of some questions assume that you have also studied Section 5.6 and can use the periodic table as a source of atomic masses.

12. What advantage does the atomic mass unit have over grams when speaking of the mass of an atom or a subatomic particle?

13. The mass of an average atom of a certain element is 6.66 times as great as the mass of an atom of carbon-12. Using either the periodic table or the table of elements, identify the element.

14. The atomic masses of the natural isotopes of neon are 19.99244 amu, 20.99395 amu, and 21.99138 amu. The average of these three masses is 20.99259 amu. The atomic mass of neon is listed as 20.1797 amu on the periodic table. Which isotope do you expect is the most abundant in nature? Explain.

Name: ______________________ Date: ______________________

ID: ______________________ Section: ______________________

15. The mass of 60.4% of the atoms of an element is 68.9257 amu. There is only one other natural isotope of that element, and its atomic mass is 70.9249 amu. Calculate the average atomic mass of the element. Using the periodic table and/or the table of the elements, write its symbol and name.

Questions 16 through 18: Percentage abundances and atomic masses (amu) of the natural isotopes of an element are given. (a) Calculate the atomic mass of each element from these data. (b) Using other information that is available to you, identify the element.

	Percentage Abundance	Atomic Mass (amu)
16.	51.82	106.9041
	48.18	108.9047
17.	57.25	120.9038
	42.75	122.9041
18.	0.193	135.907
	0.250	137.9057
	88.48	139.9053
	11.07	141.9090

Name: ______________________ Date: ______________________

ID: ______________________ Section: ______________________

Section 5.6: The Periodic Table

19. How many elements are in Period 5 of the periodic table? Write the atomic numbers of the elements in Group 3B/3.

20. Locate in the periodic table each element whose atomic number is given, and identify first the number of the period it is in and then the number of the group: (a) 20; (b) 14; (c) 43.

21. Using only a periodic table for reference, list the atomic masses of the elements whose atomic numbers are 29, 55, and 82.

22. Write the atomic masses of helium and aluminum.

Name: ______________________ Date: ______________________

ID: ______________________ Section: ______________________

Section 5.7: Elemental Symbols and the Periodic Table

23. The names, atomic numbers, or symbols of some of the elements in Figure 5.8 are given in the table that follows. Fill in the open spaces, referring only to a periodic table for any information that you need.

Table of Elements		
Name	**Atomic Number**	**Symbol**
		Mg
	8	
Phosphorus		
		Ca
Zinc		
		Li
Nitrogen		
	16	
	53	
Barium		
		K
	10	
Helium		
		Br
		Ni
Tin		
	14	

Name: ______________________ Date: ______________________

ID: ______________________ Section: ______________________

More Challenging Problems

24. Two compounds of mercury and chlorine are mercury(I) chloride and mercury(II) chloride. The amount of mercury(I) chloride that contains 71 g of chlorine has 402 g of mercury; the amount of mercury(II) chloride that has 71 g of chlorine has 201 g of mercury. Show how the Law of Multiple Proportions is illustrated by these quantities.

25. Roughly three quarters of the atoms of an element have an atomic mass of about 63 amu; the atomic mass of the remaining atoms is about 65 amu. Without calculating, estimate the atomic mass of the element to the first decimal. If you are familiar with the periodic table, locate the element and write its symbol.

26. The atomic mass of lithium on a four-significant-figure periodic table is 6.941 amu. Lithium has two natural isotopes with atomic masses of 6.10512 amu and 7.01600 amu. Calculate the percentage distribution between the two isotopes.

27. Why were scientists inclined to think of an atom as a miniature solar system in the planetary model of the atom? What are the similarities and differences between electrons in orbit around a nucleus and planets in orbit around the sun?

Name: ______________________ Date: ______________________

ID: ______________________ Section: ______________________

28. The existence of isotopes did not appear until nearly a century after Dalton proposed the atomic theory, and then it appeared in experiments more closely associated with physics than with chemistry. What does this suggest about the chemical properties of isotopes?

29. The element carbon occurs in two crystal forms, diamond and graphite. The density of the diamond form is 3.51 g/cm^3, and of graphite, 2.25 g/cm^3. The volume of a carbon atom is 1.9×10^{-24} cm^3. As stated in Section 5.5, one atomic mass unit is $\frac{1}{6.02 \times 10^{23}}$ g.

a) Calculate the density of a carbon atom.
b) Suggest a reason for the density of the atom being so much larger than the density of either form of carbon.
c) The radius of a carbon atom is roughly 1×10^5 times larger than the radius of the nucleus. What is the volume of that nucleus? (*Hint*: Volume is proportional to the cube of the radius.)
d) Calculate the density of the nucleus.
e) The radius of a period on this page is about 0.02 cm. The volume of a sphere that size is 4×10^{-5} cm^3. Calculate the mass of that sphere if it were completely filled with carbon nuclei. Express the mass in tons.

Answers to Target Checks

1. The starting arrangement of atoms, as configured as reactants before a chemical change, is rearranged to form product particles. The atoms themselves are unchanged. Since atoms are simply rearranged, their mass must be the same before and after the change.
2. The electron is the lightest, with a mass of nearly zero amu. Both the neutron and the proton are the heaviest, with a mass of about 1 amu. The neutron is slightly heavier than the proton. Electrons and protons have an electrical charge; the neutron is electrically neutral.
3. The alpha particles passed through the foil because most of the volume of an atom is empty space.
4. Isotopes of the same element cannot have the same mass number, and they must have the same atomic number (for example, ${}^{13}_{6}C$ and ${}^{14}_{6}C$ have different mass numbers and the same atomic number). Isotopes of different elements can have the same mass number, and they must have different atomic numbers (for example, ${}^{14}_{7}N$ and ${}^{14}_{6}C$ have the same mass number and different atomic numbers). Mass number is the total number of protons and neutrons in a nucleus. Atomic number is the number of protons.
5. (i) d; (ii) b

Chapter 6

Chemical Nomenclature

EXAMPLE 6.1-WB

Write the formulas of the following elements as they would be written in a chemical equation: iodine, tin, silver, oxygen, mercury.

Iodine, I_2; tin, Sn; silver, Ag; oxygen, O_2; mercury, Hg

EXAMPLE 6.2-WB

For each name below, write the formula; for each formula, write the name.

boron tribromide	________	Cl_2O	________________
trisilicon tetranitride	________	S_2O_7	________________
diphosphorus pentoxide	________	I_4O_9	________________

boron tribromide, BBr_3
trisilicon tetranitride, Si_3N_4
diphosphorus pentoxide, P_2O_5

Cl_2O, dichlorine monoxide
S_2O_7, disulfur heptoxide
I_4O_9, tetraiodine nonoxide

EXAMPLE 6.3-WB

Look only at a complete periodic table as you write (a) the names of Cl^- and Ca^{2+} and (b) the formulas of the lithium and oxide ions.

(a) chloride ion and calcium ion; (b) Li^+ and O^{2-}

As in Example 6.3 in the testbook, pay particular attention to the fact that the formula of an ion formed from a diatomic element is not necessarily itself diatomic. Oxygen occurs as a diatomic molecule, O_2, when it is uncombined, but the formula of the oxide ion is O^{2-}. O_2^{2-} does in fact exist, and it is called the peroxide ion. This illustrates the fact that you must be precise in writing chemical names and formulas.

EXAMPLE 6.4-WB

Refer only to a periodic table, if necessary, and write (a) the names of Cr^{2+} and Co^{3+} and (b) the formulas of the mercury(II) ion and the tin(II) ion.

(a) chromium(II) ion and cobalt(III) ion; (b) Hg^{2+} and Sn^{2+}

EXAMPLE 6.5-WB

(a) Write the names of Fe^{2+} and Pb^{2+}. (b) Write the formulas of silver ion and mercury(I) ion. Refer only to a periodic table.

(a) iron(II) ion and lead(II) ion; (b) Ag^{+} and Hg_2^{2+}

EXAMPLE 6.6-WB

Write the name and formula of the *-ous* acid of selenium and the name and formula of the anion formed by its total ionization.

In Example 6.6 in the textbook you determined that the formula of selenic acid is H_2SeO_4. How many oxygens does an *-ous* acid have compared with the *-ic* acid? Write the name and formula of the acid.

One fewer; H_2SeO_3, selenous acid

Now apply the system and write the anion formula and name.

SeO_3^{2-}, selenite ion

EXAMPLE 6.7-WB

(a) Write the formula of hydrobromic acid. (b) Write the name of HF.

In textbook Example 6.7 you derived the formula of hydrochloric acid, HCl. These two species follow the same logic. Complete the example.

hydrobromic acid, HBr HF, hydrofluoric acid

EXAMPLE 6.8-WB

Fill in the name and formula blanks in the following table:

Acid Name	Acid Formula	Anion Formula	Anion Name
periodic acid			
	HBrO		
		IO_2^-	

Acid Name	Acid Formula	Anion Formula	Anion Name
periodic acid	HIO_4	IO_4^-	periodate ion

Let's consider periodic acid first. (That's per-iodic, incidentally, not peri-odic, as in periodic table.) The prefix *per-*, applied to the memorized formula of chloric acid, $HClO_3$, means one more oxygen atom. Therefore, perchloric acid is $HClO_4$. Substituting iodine for chlorine, we have periodic acid as HIO_4. Remove one hydrogen ion to get the anion formula, IO_4^-, with a negative charge equal to the number of hydrogens removed from the neutral molecule. The prefix *per-* is applied to the anion name as it is to the acid name. As perchloric acid → perchlorate ion, so periodic acid → periodate ion.

If you want to reconsider your other entries, change them here on the other two lines.

Acid Name	Acid Formula	Anion Formula	Anion Name
	HBrO		
		IO_2^-	

Acid Name	Acid Formula	Anion Formula	Anion Name
periodic acid	HIO_4	IO_4^-	periodiate ion
hypobromous acid	HBrO	BrO^-	hypobromite ion
iodous acid	HIO_2	IO_2^-	iodite ion

The reasoning processes are similar for all lines in the table. In the second line there are two fewer oxygen atoms than in chloric acid, $HClO_3$, and in the third line one fewer. Prefixes and suffixes match those for the corresponding chlorine substances in Table 6.3.

EXAMPLE 6.9-WB

Fill in the name and formula blanks in the following table:

Acid Name	Acid Formula	Anion Formula	Anion Name
Hydrosulfuric acid			
		ClO^-	

Acid Name	Acid Formula	Anion Formula	Anion Name
hydrosulfuric acid	H_2S	S^{2-}	sulfide ion
hypochlorous acid	HClO	ClO^-	hypochlorite ion

We know that hydrosulfuric acid has no oxygen because of the *hydro-* and *-ic* prefix and suffix. Why is its formula H_2S rather than HS? First, sulfur is in Group 6A/16, not 7A/17. Is there another element in Group 6A/16 that forms a compound with hydrogen whose formula you know? How about oxygen and its famous hydrogen compound, H_2O, also known as water? If the compound of hydrogen and oxygen is H_2O, then the compound of hydrogen and sulfur should be H_2S. This kind of reasoning shows you how you can use the periodic table to predict names or formulas of compounds.

EXAMPLE 6.10-WB

For each of the following names, write the formula; for each formula, write the name.

periodate ion ____________ H_2SO_3 ____________

nitric acid ____________ PO_4^{3-} ____________

telluric acid ____________ SeO_3^{2-} ____________

(tellurium, Z = 52) (Se, selenium, Z = 34)

periodate ion, IO_4^- H_2SO_3, sulfurous acid
nitric acid, HNO_3 PO_4^{3-}, phosphate ion
telluric acid, H_2TeO_4 SeO_3^{2-} selenite ion

The last item in each column includes elements from Group 6A/16, the same chemical family as sulfur. The formula of telluric acid matches that of sulfuric acid, H_2SO_4. From sulfuric acid the name of SO_4^{2-} is sulfate ion. One fewer oxygen atom makes it SO_3^{2-}, sulfite ion. Substitution of selenium for sulfur in name and formula gives SeO_3^{2-}, selenite ion.

EXAMPLE 6.11-WB

Write the formulas of potassium chloride and potassium hydroxide.

These two compounds involve three ions. Begin by writing the formulas of the ions, including charges, so you can see clearly the ions that must be combined.

K^+ Cl^- OH^-

You must now decide how many cations and anions to combine for each compound so the sum of their charges is equal to zero. The potassium ion has a 1+ charge; the chloride ion has a 1– charge. How many potassium ions must combine with how many chloride ions so the sum of the charges is zero? Similarly, how many potassium ions must combine with how many hydroxide ions?

---

1 potassium ion + 1 chloride ion; 1 potassium ion + 1 hydroxide ion
In both cases, (1+) + (1–) = 0.

Now write the formulas of the two compounds, following the two steps just shown.

Potassium chloride, KCl; potassium hydroxide, KOH
There are no subscripts in these formulas because each ion appears only once.

EXAMPLE 6.12-WB

Write the formulas of ammonium nitrate, ammonium carbonate, and calcium carbonate.

This question introduces you to two new ions, ammonium ion and carbonate ion. Both are polyatomic ions. See if you can go all the way to the formulas this time. Remember that you are combining formulas of *ions*, not symbols of atoms. Keep the distinct identity of each ion and use parentheses where needed.

Ammonium nitrate, NH_4NO_3; ammonium carbonate, $(NH_4)_2CO_3$;
calcium carbonate, $CaCO_3$

The ions: NH_4^+ Ca^{2+} NO_3^- CO_3^{2-}

The number of cation and anions: 1 NH_4^+ + 1 NO_3^- give $(1+) + (1-) = 0$

2 NH_4^+ + 1 CO_3^{2-} give $2 \times (1+) + (2-) = 0$

1 Ca^{2+} + 1 CO_3^{2-} give $(2+) + (2-) = 0$

Notice that ammonium and nitrate ions keep their identities in NH_4NO_3. The nitrogens in the two ions are not combined.

EXAMPLE 6.13-WB

Write the formulas of copper(II) chloride, nickel nitrate, and magnesium hydrogen sulfate.

Copper(II) chloride, $CuCl_2$; nickel nitrate, $Ni(NO_3)_2$;
magnesium hydrogen sulfate, $Mg(HSO_4)_2$

The (II) in copper(II) chloride indicates the copper ion with a 2+ charge, so the ions are Cu^{2+} and Cl^-. The nickel ion is one of the three transition-metal ions whose charge must be memorized because they have only one common charge. In this case we have Ni^{2+}. It is combined with a nitrate ion, NO_3^-. The hydrogen sulfate ion is sulfate ion (SO_4^{2-}) with a hydrogen ion (H^+) attached: HSO_4^-. That is combined with a magnesium ion, Mg^{2+}.

EXAMPLE 6.14-WB

Write the name of each of the following compounds:

$NaHSO_3$ ____________________

K_2HPO_4 ____________________

$ZnCO_3$ ____________________

HgS ____________________

$(NH_4)_2SeO_4$ ____________________

(selenium, Z = 34)

$NaHSO_3$, sodium hydrogen sulfite
K_2HPO_4, potassium hydrogen phosphate
$ZnCO_3$, zinc carbonate
HgS, mercury(II) sulfide
$(NH_4)_2SeO_4$, ammonium selenate

One 2– charge from the sulfide ion in HgS must be balanced by 2+ from a mercury(II) ion. The zinc ion in $ZnCO_3$ has only one common charge, so its charge is not indicated in its name even though it is a transition element ion. In $(NH_4)_2SeO_4$, selenium substitutes for its family member, sulfur, in sulfate ion, SO_4^{2-}, so SeO_4^{2-} is the selenate ion.

EXAMPLE 6.15-WB

a) How many water molecules are associated with each formula unit of anhydrous sodium carbonate in $CoSO_4 \cdot 7\ H_2O$? Name the hydrate.
b) Write the formula of barium perchlorate trihydrate.

---

a) Seven; cobalt(II) sulfate heptahydrate b) $Ba(ClO_4)_2 \cdot 3\ H_2O$

Name: ______________________ Date: ______________________

ID: ______________________ Section: ______________________

Questions, Exercises, and Problems

General Instructions: *Most of the questions in this chapter ask that you write the name of any species if the formula is given, or the formula if the name is given. You will be reminded of this briefly at the beginning of each such block of questions. You should try to follow these instructions without reference to anything except a clean periodic table, one that has nothing written on it. Names and/or atomic numbers are given in questions involving elements not shown in Figure 5.8. An asterisk (*) marks a substance containing an ion you are not expected to recognize. If you cannot predict what it is from the periodic table, refer to Table 6.7 or 6.8.*

Section 6.2: Formulas of Elements

1. The elements of Group 8A/18 are stable as monatomic atoms. Write their formulas.

Questions 2 to 4: Given names, write formulas; given formulas, write names.

2. Fluorine, boron, nickel, sulfur

3. Cr, Cl_2, Be, Fe

4. Krypton, copper, manganese, nitrogen

Name: ______________________ Date: ______________________

ID: ______________________ Section: ______________________

Section 6.3: Compounds Made from Two Nonmetals

5. Given names, write formulas; given formulas, write names:
Dichlorine oxide, tribromine octoxide, HBr(g), P_2O_3

Section 6.4: Names and Formulas of Ions Formed by One Element

6. Explain how monatomic anions are formed from atoms, in terms of protons and electrons.

Questions 7 and 8: Given names, write formulas; given formulas, write names.

7. Cu^+, I^-, K^+, Hg_2^{2+}, S^{2-}

8. Iron(III) ion, hydrogen ion, oxide ion, aluminum ion, barium ion

Section 6.5: Acids and the Anions Derived from Their Total Ionization

9. How do you recognize the formula of an acid?

10. How many ionizable hydrogens are in monoprotic, diprotic, and triprotic acids?

Name: ______________________ Date: ______________________

ID: ______________________ Section: ______________________

11. The following table has spaces for the names and formulas covered by Goal 5 that are not in textbook Question 11. Fill in the remaining blanks. (Selenium, Se, Z = 34; tellurium, Te, Z = 54).

Acid Name	Acid Formula	Ion Name	Ion Formula
	H_2SO_4		
		Carbonate	
			ClO_3^-
Hydrofluoric			
	$HBrO_3$		
		Sulfite	
			AsO_4^{3-}
Periodic			
	H_2SeO_3		
		Tellurite	
			IO^-
Hypobromous			
	H_2TeO_4		
		Perbromate	
			Br^-

Section 6.6: Names and Formulas of Acid Anions

12. Explain how an anion can behave like an acid. Is it possible for a cation to be an acid?

Name: ______________________ Date: ______________________

ID: ______________________ Section: ______________________

Questions 13 and 14: Given names, write formulas; given formulas, write names.

13. Hydrogen sulfite ion, hydrogen carbonate ion

14. $HSeO_3^-$, HTe^-

Section 6.7: Names and Formulas of Other Acids and Ions

Questions 15 and 16: Given names, write formulas; given formulas, write names. Refer to Table 6.7 or 6.8 only if necessary.

15. NH_4^+, CN^-

16. Hydroxide ion, cadmium ion (Z = 48)

Section 6.8: Formulas of Ionic Compounds

Write the formulas of the compounds in Questions 17 to 19.

17. Calcium hydroxide, ammonium bromide, potassium sulfate

18. Magnesium oxide, aluminum phosphate, sodium sulfate, calcium sulfide

19. Barium sulfite, chromium(III) oxide, potassium periodate, calcium hydrogen phosphate

Name: ______________________ Date: ______________________

ID: ______________________ Section: ______________________

Section 6.9: Names of Ionic Compounds

Name the compounds in Questions 20 to 22.

20. Li_3PO_4, $MgCO_3$, $Ba(NO_3)_2$

21. KF, NaOH, CaI_2, $Al_2(CO_3)_3$

22. $CuSO_4$, $Cr(OH)_3$, Hg_2I_2

Section 6.10: Hydrates

23. Among the following, identify all hydrates and anhydrous compounds:
$NiSO_4 \cdot 6\ H_2O$, KCl, $Na_3PO_4 \cdot 12\ H_2O$.

24. Epsom salt has the formula $MgSO_4 \cdot 7\ H_2O$. How many water molecules are associated with one formula unit of $MgSO_4$? Write the chemical name of Epsom salt.

25. Write the formulas of ammonium phosphate trihydrate and potassium sulfide pentahydrate.

Name: ______________________ Date: ______________________

ID: ______________________ Section: ______________________

Questions 26 to 40: Items in the remaining questions are selected at random from various sections of the unit. Unless marked with an asterisk (*), all names and formulas are included in the Goals and should be found with reference to no more than a periodic table. Ions in compounds marked with an asterisk are included in Tables 6.7 and 6.8, or, if the unfamiliar ion is monatomic, the atomic number of the element is given. In all questions, given a name, write the formula; given a formula, write the name.

26. Perchlorate ion, barium carbonate, NH_4I, PCl_3

27. HS^-, $BeBr_2$, aluminum nitrate, oxygen difluoride

28. Mercury(I) ion, cobalt(II) chloride, SiO_2, $LiNO_2$

29. N^{3-}, $Ca(ClO_3)_2$, iron(III) sulfate, phosphorus pentachloride

30. Tin(II) fluoride, potassium chromate*, LiH, $FeCO_3$

31. HNO_2, $Zn(HSO_4)_2$, potassium cyanide*, copper(I) fluoride

32. Magnesium nitride, lithium bromite, $NaHSO_3$, KSCN*

Name: ______________________ Date: ______________________

ID: ______________________ Section: ______________________

33. $Ni(HCO_3)_2$, CuS, chromium(III) iodate, potassium hydrogen phosphate

34. Selenium dioxide (selenium, Z = 34), magnesium nitrite, $FeBr_2$, Ag_2O

35. SnO, $(NH_4)_2Cr_2O_7$*, sodium hydride*, oxalic acid*

36. Cobalt(III) sulfate, iron(III) iodide, $Cu_3(PO_4)_2$, $Mn(OH)_2$

37. Al_2Se_3 (Se is selenium, Z = 34), $MgHPO_4$, potassium perchlorate, bromous acid

38. Strontium iodate (strontium, Z = 38), sodium hypochlorite, Rb_2SO_4, P_2O_5

39. ICl, $AgC_2H_3O_2$*, lead(II) dihydrogen phosphate, gallium fluoride (gallium, Z = 31)

40. Magnesium sulfate, mercury(II) bromite, $Na_2C_2O_4$*, $Mn(OH)_3$

Chapter 7

Chemical Formula Relationships

EXAMPLE 7.1-WB

(a) What are the number of atoms of each element in one formula unit of aluminum perchlorate?
(b) Write the formula for a compound if its formula unit contains one sulfur atom and three oxygen atoms.

Before you can answer Part (a) you need the formula of aluminum perchlorate.

$Al(ClO_4)_3$

Aluminum is in Group 3A/13, so its ion has a 3+ charge, and the formula of the aluminum ion is therefore Al^{3+}. The perchlorate ion is derived from the memorized chloric acid, $HClO_3$. The *per-* prefix indicates one more oxygen than in the *-ic* acid, so perchloric acid is $HClO_4$. To change from an *-ic* acid to its corresponding ion, change *-ic* to *-ate:* ClO_4^- is the perchlorate ion.

Now you have the formula, so complete Part (a).

---

aluminum perchlorate: 1 aluminum atom, 3 chlorine atoms, 12 oxygen atoms

The three perchlorate ions each contain four oxygen atoms, $3 \times 4 = 12$ oxygen atoms.

Complete Part (b).

SO_3

You may have been tempted to write a subscript 1 after the symbol for the sulfur atom, but remember that 1's are not included in chemical formulas.

EXAMPLE 7.2-WB

Calculate the formula mass of (a) magnesium sulfate and (b) aluminum sulfide.

First you need the formulas. What are they?

$MgSO_4$ and Al_2S_3

Mg^{2+} and SO_4^{2-} combine on a 1:1 ratio. With aluminum sulfide, it takes three 2– charges from S^{2-} to balance two 3+ charges from Al^{3+}.

Work on $MgSO_4$ first. Using the preceding formula and the periodic table for atomic masses, write a horizontal setup for the problem, but do not calculate the answer yet.

1 Mg atom + 1 S atom + 4 O atoms = 24.31 amu + 32.07 amu + 4(16.00 amu)

Now use your calculator to find the sum. Do it, if you can, without writing any other numbers, just the final answer.

$MgSO_4$: 24.31 amu + 32.07 amu + 4(16.00 amu) = 120.38 amu

Next write the horizontal setup for Al_2S_3 and find the formula mass on your calculator.

Al_2S_3: 2(26.98 amu) + 3(32.07 amu) = 150.17 amu

EXAMPLE 7.3-WB

Calculate the formula mass of (a) mercury(II) sulfide and (b) nickel carbonate.

Begin with the formulas of these two compounds.

HgS and $NiCO_3$

One sulfide ion, S^{2-}, is needed to balance the 2+ charge of a mercury(II) ion, Hg^{2+}. The nickel ion, Ni^{2+}, is one of the transition element ions that has a charge that must be memorized. The carbonate ion, CO_3^{2-}, is derived from the memorized carbonic acid, H_2CO_3.

Calculate the formula mass of HgS. Consider the significant figures carefully.

200.6 amu + 32.07 amu = 232.7 amu
↑ ↑ ↑
1 decimal place 2 decimal places 1 decimal place

Now determine the formula mass of $NiCO_3$.

58.69 amu + 12.01 amu + 3(16.00 amu) = 118.70 amu

In this case, you must include the final zero even though it does not appear on your calculator. All atomic masses are known to two decimal places, so their sum is also known to two decimal places.

EXAMPLE 7.4-WB

Fluorine is the most reactive of all the elements. Adrenaline is the molecule that triggers the "fight or flight" reflex in us; its formula is $C_9H_{13}NO_3$. Calculate the molecular masses of these substances.

Use what you learned from textbook Example 7.4 as you complete this example.

F_2: 2(19.00 amu) = 38.00 amu

$C_9H_{13}NO_3$: 9(12.01 amu) + 13(1.008 amu) + 14.01 amu + 3(16.00 amu) = 183.20 amu

Unless you are specifically told otherwise, always assume that a diatomic elemental name refers to the natural form of the element at normal temperatures. Thus a question or statement about fluorine refers to F_2, not F.

EXAMPLE 7.5-WB

(a) What does one mole of carbon atoms have in common with one mole of oxygen atoms?
(b) How many nitrogen molecules are in 7.08 moles of nitrogen?

Consider Part (a) first. Review the definition of the mole before you write your answer.

One mole of carbon atoms has the same number of atoms as one mole of oxygen atoms.

This is similar to the fact that one dozen eggs has the same number of objects as one dozen golf balls.

Write the *GIVEN, WANTED,* and *PER/PATH* for Part (b), and then complete the setup and solution.

GIVEN: 7.08 mol N_2 *WANTED:* N_2 molecules

PER/PATH: mol N_2 $\xrightarrow{6.02 \times 10^{23}\ N_2\ \text{molecules/mol}\ N_2}$ N_2 molecules

$$7.08\ \text{mol}\ N_2 \times \frac{6.02 \times 10^{23}\ N_2\ \text{molecules}}{\text{mol}\ N_2} = 4.26 \times 10^{24}\ N_2\ \text{molecules}$$

EXAMPLE 7.6-WB

(a) What does 12.01 g of carbon atoms have in common with 16.00 g of oxygen atoms?
(b) The molecular mass of the explosive TNT is 227 amu. What is the molar mass of TNT?
(c) Determine the molar masses of carbon, oxygen, and carbon monoxide.

Begin by writing your answer to Part (a).

12.0 g C atoms (12.0 g C/mol C) is the same number of atoms as 16.0 g O atoms (16.0 g O/mol O).

Compare Part (a) of this example with Part (a) in Example 7.5-WB. Notice that we essentially asked the same question in two different ways.

Part (b) is next.

---

227 g/mol

Molar mass in g/mol is numerically equal to molecular mass in amu.

Finish up the example by completing Part (c).

C: 12.01 g/mol C; O_2: 2(16.00 g/mol O) = 32.00 g/mol O_2;
CO: 12.01 g/mol C + 16.00 g/mol O = 28.01 g/mol CO

EXAMPLE 7.7-WB

How many moles of aluminum sulfate are in 132 g of the compound? (Among the many uses of aluminum sulfate are tanning leather, fireproofing and waterproofing cloth, treating sewage, and making antiperspirants.)

The mass → mole conversion is by molar mass. Begin with the formula of aluminum sulfate.

$Al_2(SO_4)_3$

The formula requires "two of the 3+'s" (two Al^{3+}) to balance "three of the 2–'s" (three SO_4^{2-}) at 6+ and 6–, as with aluminum sulfide in Example 7.2-WB.

Now you need the molar mass.

2(26.98 g/mol Al) + 3(32.07 g/mol S) + 12(16.00 g/mol O) = 342.17 g/mol $Al_2(SO_4)_3$

Set up and solve the problem.

GIVEN: 132 g $Al_2(SO_4)_3$ *WANTED*: mol $Al_2(SO_4)_3$

PER/PATH: g $Al_2(SO_4)_3$ $\xrightarrow{342.17 \text{ g } Al_2(SO_4)_3/\text{mol } Al_2(SO_4)_3}$ mol $Al_2(SO_4)_3$

$$132 \text{ g } Al_2(SO_4)_3 \times \frac{1 \text{ mol } Al_2(SO_4)_3}{342.17 \text{ g } Al_2(SO_4)_3} = 0.386 \text{ mol } Al_2(SO_4)_3$$

EXAMPLE 7.8-WB

Ammonia is an important industrial chemical that is used in refrigerants and in the manufacture of nitric acid, explosives, synthetic fibers, and fertilizers. What is the mass of one billion billion ($1.00 \times 10^9 \times 10^9 = 1.00 \times 10^{18}$) molecules of ammonia?

First, write the formula of ammonia.

NH_3

Now take it all the way to the answer.

GIVEN: 1.00×10^{18} molecules NH_3 *WANTED*: g NH_3

PER/PATH: molecules NH_3 $\xrightarrow{6.02 \times 10^{23} \text{ molecules } NH_3\text{/mol } NH_3}$ mol NH_3 $\xrightarrow{17.03 \text{ g } NH_3\text{/mol } NH_3}$ g NH_3

$$1.00 \times 10^{18} \text{ molecules } NH_3 \times \frac{1 \text{ mol } NH_3}{6.02 \times 10^{23} \text{ molecules } NH_3} \times \frac{17.03 \text{ g } NH_3}{\text{mol } NH_3} = 2.83 \times 10^{-5} \text{ g } NH_3$$

This very small mass, about 6/100,000,000 of a pound, suggests again the enormous number of molecules in a mole.

EXAMPLE 7.9-WB

Vitamin C is probably the most controversial vitamin because of debates over its effectiveness in preventing and curing the common cold and in prevention of cancer. Its molecular formula is $C_6H_8O_6$. What is the percentage of carbon in vitamin C?

This is a straightforward percentage problem after you have determined the molar mass of the compound.

GIVEN: 6(12.01 g C); 176.12 g $C_6H_8O_6$ *WANTED*: % C

EQUATION: $$\% \text{ C} = \frac{\text{g C}}{\text{g } C_6H_8O_6} \times 100 = \frac{6(12.01) \text{ g C}}{176.12 \text{ g } C_6H_8O_6} \times 100 = 40.92\% \text{ C}$$

EXAMPLE 7.10-WB

Determine the percentage composition of vitamin C, $C_6H_8O_6$.

In Example 7.9-WB, you calculated the molar mass of vitamin C as 176.12 g/mol and its percentage carbon as 40.92%. Determine the percentage hydrogen and percentage oxygen, and check your result.

GIVEN: 8(1.008 g H); 6(16.00 g O); 176.12 g $C_6H_8O_6$ *WANTED:* % H, % O

EQUATION: $\% \text{ H} = \frac{\text{g H}}{\text{g } C_6H_8O_6} \times 100 = \frac{8(1.008 \text{ g H})}{176.12 \text{ g } C_6H_8O_6} \times 100 = 4.579\% \text{ H}$

$\% \text{ O} = \frac{\text{g O}}{\text{g } C_6H_8O_6} \times 100 = \frac{6(16.00) \text{ g O}}{176.12 \text{ g } C_6H_8O_6} \times 100 = 54.51\% \text{ O}$

CHECK: 40.92% + 4.579% + 54.51% = 100.01%

EXAMPLE 7.11-WB

An experiment requires that enough calcium fluoride be used to yield 1.91 g of calcium. How much calcium fluoride must be weighed out?

You know from the text at the beginning of textbook Section 7.6 that 51.33% of a sample of CaF_2 is calcium. Complete the problem.

GIVEN: 1.91 g Ca *WANTED:* g CaF_2

PER/PATH: g Ca $\xrightarrow{51.33 \text{ g Ca}/100 \text{ g } CaF_2}$ g CaF_2

$1.91 \text{ g Ca} \times \frac{100 \text{ g } CaF_2}{51.33 \text{ g Ca}} = 3.72 \text{ g } CaF_2$

EXAMPLE 7.12-WB

Determine the number of milligrams of oxygen in a sample of vitamin C, $C_6H_8O_6$, that contains 88.7 mg of carbon.

PLAN your solution and go all the way to the final answer. Any ratio that you can set up as a conversion factor with grams in the numerator and denominator can also be set up as an equivalent *PER* expression with milligrams in the top and bottom of the fraction.

GIVEN: 88.7 mg C *WANTED:* mg O

PER/PATH: mg C $\xrightarrow{6(16.00)\text{ mg O}/6(12.01)\text{ mg C}}$ mg O

$$88.7\text{ mg C} \times \frac{6(16.00)\text{ mg O}}{6(12.01)\text{ mg C}} = 118\text{ mg O}$$

EXAMPLE 7.13-WB

Acetylcholine, $C_7H_{16}NO_2^+$, is a neurotransmitter, a compound that enables communication between nerve cells or between a nerve cell and another cell. (a) How many atoms of each element are in acetylcholine (Sec. 7.1)? (b) What is the mass of one unit of $C_7H_{16}NO_2^+$ (Sec. 7.2)? (c) When a nerve cell delivers acetylcholine, 3.0×10^3 cations are typically released from each small storage pocket, known as a vesicle. How many moles of acetylcholine is this (Sec. 7.3)? (d) What is the molar mass of acetylcholine (Sec. 7.4)? (e) What is the mass (in grams) of 3.0×10^3 acetylcholine cations (Sec. 7.5)? (f) What is the percentage composition of acetylcholine (Sec. 7.6)?

This example is a summary and review of Chapter 7 thus far. Each part of the question has a reference to the appropriate section for review, as necessary. We'll guide you in working this problem one part at a time. If you have trouble with any part, review the associated section before moving on to the next part of the question.

Begin with Part (a).

---

7 carbon atoms, 16 hydrogen atoms, 1 nitrogen atom, 2 oxygen atoms

Now complete Part (b).

7(12.01 amu) + 16(1.008 amu) + 14.01 amu + 2(16.00 amu) = 146.21 amu

The mass of an individual particle is typically expressed in amu's. If you calculated the mass in g/mol, you determined the mass of *one mole* of particles rather than the mass of *one* particle.

Part (c) is next. We'll show the complete *PLAN* and solution, but if you can answer the question without the *PLAN*, do so.

GIVEN: 3.0×10^3 units $C_7H_{16}NO_2^+$ *WANTED:* mol $C_7H_{16}NO_2^+$

PER/PATH: units $C_7H_{16}NO_2^+$ $\xrightarrow{1 \text{ mol } C_7H_{16}NO_2^+/6.02 \times 10^{23} \text{ units } C_7H_{16}NO_2^+}$ mol $C_7H_{16}NO_2^+$

$$3.0 \times 10^3 \text{ units } C_7H_{16}NO_2^+ \times \frac{1 \text{ mol } C_7H_{16}NO_2^+}{6.02 \times 10^{23} \text{ units } C_7H_{16}NO_2^+} = 4.98 \times 10^{-21} \text{ mol } C_7H_{16}NO_2^+$$

Part (d) asks for the molar mass of acetylcholine. Don't forget that you determined its formula mass in Part (b).

146.21 g/mol

To convert from the mass of a particle to the mass of a mole of particles, change the units.

Part (e) is made simpler by using your results from Parts (c) and (d).

GIVEN: 4.98×10^{-21} mol $C_7H_{16}NO_2^+$ *WANTED:* g $C_7H_{16}NO_2^+$

PER/PATH: mol $C_7H_{16}NO_2^+$ $\xrightarrow{146.21 \text{ g } C_7H_{16}NO_2^+/\text{mol } C_7H_{16}NO_2^+}$ g $C_7H_{16}NO_2^+$

$$4.98 \times 10^{-21} \text{ mol } C_7H_{16}NO_2^+ \times \frac{146.21 \text{ g } C_7H_{16}NO_2^+}{\text{mol } C_7H_{16}NO_2^+} = 7.28 \times 10^{-19} \text{ g } C_7H_{16}NO_2^+$$

Wrap up this example with Part (f), determininination of percentage composition.

$$\frac{7(12.01)\text{ g C}}{146.21\text{ g }C_7H_{16}NO_2^+} \times 100 = 57.50\%\text{ C} \qquad \frac{16(1.008)\text{ g H}}{146.21\text{ g }C_7H_{16}NO_2^+} \times 100 = 11.03\%\text{ H}$$

$$\frac{14.01\text{ g N}}{146.21\text{ g }C_7H_{16}NO_2^+} \times 100 = 9.582\%\text{ N} \qquad \frac{2(16.00)\text{ g O}}{146.21\text{ g }C_7H_{16}NO_2^+} \times 100 = 21.89\%\text{ O}$$

57.50% + 11.03 % + 9.582% + 21.89% = 100.00%

EXAMPLE 7.14-WB

Write EF after each formula that is an empirical formula. Write the empirical formula after each compound whose formula is not already an empirical formula.

C_2H_6: __________ C_3H_6: __________ C_3H_8O: __________ CH_5N: __________

C_2H_6: CH_3 C_3H_6: CH_2 C_3H_8O: EF CH_5N: EF

EXAMPLE 7.15-WB

An oxide of sulfur is 40.1% sulfur and 59.9% oxygen by mass. Find the empirical formula of this compound.

As we stated in the textbook, percentage composition represents the grams of each element in a 100-g sample. If you had 100 g of this compound, you would have 40.1 g S and 59.9 g O. These quantities go into the grams column of the empirical formula table. Take the table all the way to moles.

Element	Grams	Moles	Mole Ratio	Formula Ratio	Empirical Formula

---

Element	Grams	Moles	Mole Ratio	Formula Ratio	Empirical Formula
S	40.1	1.25			
O	59.9	3.74			

40.1 g S ÷ 32.07 g/mol S = 1.25 mol S 59.9 g O ÷ 16.00 g/mol O = 3.74 mol O

Recall that to go from moles to mole ratio, divide by the smallest number of moles. Round to whole numbers to get formula ratio numbers. Finally, use the formula ratio numbers as subscripts for the empirical formula. Finish the problem.

Element	Grams	Moles	Mole Ratio	Formula Ratio	Empirical Formula
S	40.1	1.25	1.00	1	SO_3
O	59.9	3.74	2.99	3	

1.25 mol ÷ 1.25 mol = 1.00   3.74 mol ÷ 1.25 mol = 2.99

EXAMPLE 7.16-WB

A 1.04-g sample known to be a pure oxide of chromium is analyzed and found to have 0.710 g chromium. What is the empirical formula of the compound?

Begin by determining the mass of oxygen in the sample.

1.04 g chromium oxide − 0.710 g chromium = 0.33 g oxygen

You can now start your empirical formula table. Go all the way to the mole ratio.

Element	Grams	Moles	Mole Ratio	Formula Ratio	Empirical Formula

Element	Grams	Moles	Mole Ratio	Formula Ratio	Empirical Formula
Cr	0.710	0.0137	1.00		
O	0.33	0.021	1.5		

0.170 g Cr ÷ 52.00 g/mol Cr = 0.0137 mol Cr   0.33 g O ÷ 16.00 g/mol O = 0.021 mol O
0.0137 mol ÷ 0.0137 mol = 1.00   0.021 mol ÷ 0.0137 mol = 1.5

The mole ratio is 1.5/1 or $1\frac{1}{2}/1$ or $\frac{3}{2}/1$. You need to think about how to change this ratio to a ratio of whole numbers. Since two is in the denominator of the fraction, multiplication by 2 will cancel the denominator and give you a ratio of whole numbers. This type of thinking will help you in general in changing from mole ratios to formula ratios. Complete the remainder of the problem by filling in the preceding table.

Element	Grams	Moles	Mole Ratio	Formula Ratio	Empirical Formula
Cr	0.710	0.0137	1.00	2	Cr_2O_3
O	0.33	0.021	1.5	3	

1.00 × 2 = 2   1.5 × 2 = 3

EXAMPLE 7.17-WB

Determine the empirical formula of a compound that contains 68.8% carbon, 4.96% hydrogen, and 26.2% oxygen.

Try to get the complete setup and solution on your own.

Element	Grams	Moles	Mole Ratio	Formula Ratio	Empirical Formula
C	68.8	5.73	3.49	7	
H	4.96	4.92	3.00	6	$C_7H_6O_2$
O	26.2	1.64	1.00	2	

To convert from grams to moles:
68.8 g C ÷ 12.01 g/mol C = 5.73 mol C   4.96 g H ÷ 1.008 g/mol H = 4.92 mol H
26.2 g O ÷ 16.00 g/mol O = 1.64 mol O
To convert from moles to mole ratio:
5.73 mol ÷ 1.64 mol = 3.49   4.92 mol ÷ 1.64 mol = 3.00   1.64 mol ÷ 1.64 mol = 1.00
Multiplication of the mole ratio figures by 2 yields integers for the formula ratio column:
3.49 × 2 = 6.98 or 7   3.00 × 2 = 6   1.00 × 2 = 2

EXAMPLE 7.18-WB

Acetaminophen is a commonly used non-aspirin pain reliever and fever reducer. It has the percentage composition 63.56% carbon, 6.00% hydrogen, 9.27% nitrogen, and 21.17% oxygen. It has a molar mass of 151 g/mol. Determine the molecular formula of acetaminophen.

The problem statement does not ask for the empirical formula, but finding it is a necessary intermediate step in finding the molecular formula.

Element	Grams	Moles	Mole Ratio	Formula Ratio	Empirical Formula
C	63.56	5.292	7.99	8	
H	6.00	5.95	8.99	9	$C_8H_9NO_2$
N	9.27	0.662	1.00	1	
O	21.17	1.323	2.00	2	

Now your task is to find the number of empirical formula units per molecule of acetaminophen. This requires the molar mass of the empirical formula unit.

8(12.01 g/mol C) + 9(1.008 g/mol H) + 14.01 g/mol N + 2(16.00 g/mol O) = 151.16 g/mol $C_8H_9NO_2$

To complete the problem, write the molecular formula.

$C_8H_9NO_2$

The molar mass of the empirical formula unit is the same as the molar mass of the molecule, so the empirical formula and the molecular formula are the same.

Name: ______________________ Date: ______________________

ID: ______________________ Section: ______________________

Questions, Exercises, and Problems

Section 7.1: The Number of Atoms in a Formula

1. How many atoms of each element are in a formula unit of aluminum nitrate?

Section 7.2: Molecular Mass and Formula Mass

2. Why is it proper to speak of the molecular mass of water but not of the molecular mass of sodium nitrate?

3. Which of the three terms, *atomic mass*, *molecular mass*, or *formula mass*, is most appropriate for each of the following: ammonia, calcium oxide, barium, chlorine, sodium carbonate?

4. Find the formula mass of each of the following substances:
a) Lithium chloride

b) Aluminum carbonate

c) Ammonium sulfate

d) Butane, C_4H_{10} (molecular mass)

Name: ______________________ Date: ______________________

ID: ______________________ Section: ______________________

e) Silver nitrate

f) Manganese(IV) oxide

g) Zinc phosphate

Section 7.3: The Mole Concept

5. What do quantities representing one mole of iron atoms and one mole of ammonia molecules have in common?

6. Is the mole a number? Explain.

7. Determine how many atoms or molecules are in each of the following:
a) 7.75 moles of methane, CH_4

Answer: [ ]

Name: ______________________ Date: ______________________

ID: ______________________ Section: ______________________

b) 0.0888 mole of carbon monoxide

Answer: [ ]

c) 57.8 moles of iron

Answer: [ ]

8. Calculate the number of moles in each of the following:
a) 2.45×10^{23} acetylene molecules, C_2H_2

Answer: [ ]

b) 6.96×10^{24} sodium atoms

Answer: [ ]

Name: ______________________ Date: ______________________

ID: ______________________ Section: ______________________

Section 7.4: Molar Mass

9. In what way are the molar mass of atoms and atomic mass the same?

10. Find the molar mass of all the following substances.

a) C_3H_8

Answer: [ ]

b) C_6Cl_5OH

Answer: [ ]

c) Nickel phosphate

Answer: [ ]

Name: ______________________ Date: ______________________

ID: ______________________ Section: ______________________

d) Zinc nitrate

Answer: [ ]

Section 7.5: Conversion among Mass, Number of Moles, and Number of Units

Questions 11 and 12: Find the number of moles for each mass of substance given.

11. a) 6.79 g oxygen

Answer: [ ]

b) 9.05 g magnesium nitrate

Answer: [ ]

c) 0.770 g aluminum oxide

Answer: [ ]

Name: ____________________ Date: ____________________

ID: ____________________ Section: ____________________

d) 659 g C_2H_5OH

Answer: [ ]

e) 0.394 g ammonium carbonate

Answer: [ ]

f) 34.0 g lithium sulfide

Answer: [ ]

12. a) 0.797 g potassium iodate

Answer: [ ]

Name: ____________________ Date: ____________________

ID: ____________________ Section: ____________________

b) 68.6 g beryllium chloride

Answer: [ ]

c) 302 g nickel nitrate

Answer: [ ]

Questions 13 and 14: Calculate the mass of each substance from the number of moles given.

13. a) 0.769 mol lithium chloride

Answer: [ ]

b) 57.1 mol acetic acid, $HC_2H_3O_2$

Answer: [ ]

Name: ______________________ Date: ______________________

ID: ______________________ Section: ______________________

c) 0.68 mol lithium

Answer: [ ]

d) 0.532 mol iron(III) sulfate

Answer: [ ]

e) 8.26 mol sodium acetate (acetate ion, $C_2H_3O_2^-$)

Answer: [ ]

14. a) 0.379 mol lithium sulfate

Answer: [ ]

b) 4.82 mol potassium oxalate (oxalate ion, $C_2O_4^{2-}$)

Answer: [ ]

Name: ______________________ Date: ______________________

ID: ______________________ Section: ______________________

c) 0.132 mol lead(II) nitrate

Answer: [ ]

Questions 15 and 16: Calculate the number of atoms, molecules, or formula units that are in each given mass.

15. a) 29.6 g lithium nitrate

Answer: [ ]

b) 0.151 g lithium sulfide

Answer: [ ]

c) 457 g iron(III) sulfate

Answer: [ ]

Name: ____________________ Date: ____________________

ID: ____________________ Section: ____________________

16. a) 0.0023 g iodine molecules

Answer: [ ]

b) 114 g $C_2H_4(OH)_2$

Answer: [ ]

c) 9.81 g chromium(III) sulfate

Answer: [ ]

17. Calculate the mass of each of the following:
a) 4.30×10^{21} molecules of $C_{19}H_{37}COOH$

Answer: [ ]

b) 8.67×10^{24} atoms of fluorine

Answer: [ ]

Name: ______________________ Date: ______________________

ID: ______________________ Section: ______________________

c) 7.23×10^{23} formula units of nickel chloride

Answer: [ ]

18. On a certain day the financial pages quoted the price of gold at $478 per troy ounce (1 troy ounce = 31.1 g). What is the price of a single atom of gold (Z = 79)?

Answer: [ ]

19. One who sweetens coffee with two teaspoons of sugar, $C_{12}H_{22}O_{11}$, uses about 0.65 g. How many sugar molecules is this?

Answer: [ ]

Name: ______________________ Date: ______________________

ID: ______________________ Section: ______________________

The stable form of certain elements is as a two-atom molecule. Fluorine and nitrogen are two of those elements. As you answer Question 20, keep in mind that the chemical formula of a molecule identifies precisely what is in the individual molecule.

20. a) What is the mass of 4.12×10^{24} N atoms?

Answer: [ ]

b) What is the mass of 4.12×10^{24} N_2 molecules?

Answer: [ ]

c) How many atoms are in 4.12 g N?

Answer: [ ]

d) How many molecules are in 4.12 g N_2?

Answer: [ ]

Name: ______________________ Date: ______________________

ID: ______________________ Section: ______________________

e) How many atoms are in 4.12 g N_2?

Answer: [ ]

Section 7.6: Mass Relationships among Elements in a Compound: Percentage Composition

21. Calculate the percentage composition of each compound:
a) Ammonium nitrate

Answer: [ ]

b) Aluminum sulfate

Answer: [ ]

Name: ______________________ Date: ______________________

ID: ______________________ Section: ______________________

c) Ammonium carbonate

Answer: [ ]

d) Calcium oxide

Answer: [ ]

e) Manganese(IV) sulfide

Answer: [ ]

Name: ______________________ Date: ______________________

ID: ______________________ Section: ______________________

22. Lithium fluoride is used as a flux when welding or soldering aluminum. How many grams of lithium are in 1.00 lb (454 g) of lithium fluoride?

Answer: [ ]

23. Potassium sulfate is found in some fertilizers as a source of potassium. How many grams of potassium can be obtained from 57.4 g of the compound?

Answer: [ ]

24. Zinc cyanide, $Zn(CN)_2$, is an important compound in zinc electroplating. How many grams of the compound must be dissolved in a test bath in a laboratory to introduce 146 g of zinc into the solution?

Answer: [ ]

25. Molybdenum ($Z = 42$) is an important element in making steel alloys. It comes from an ore called wulfenite, $PbMoO_4$. What mass of pure wulfenite must be treated to obtain 201 kg Mo?

Answer: [ ]

Name: ______________________ Date: ______________________

ID: ______________________ Section: ______________________

26. How many grams of the insecticide calcium chlorate must be measured if a sample is to contain 4.17 g chlorine?

Answer: [ ]

Section 7.8: Empirical Formula of a Compound

27. Explain why C_6H_{10} must be a molecular formula, while C_7H_{10} could be a molecular formula, an empirical formula, or both.

28. A certain compound is 52.2% carbon, 13.0% hydrogen, and 34.8% oxygen. Find the empirical formula of the compound.

Answer: [ ]

Name: ______________________ Date: ______________________

ID: ______________________ Section: ______________________

29. 11.89 g of iron is exposed to a stream of oxygen until it reacts to produce 16.99 g of a pure oxide of iron. What is the empirical formula of the product?

Answer: [ ]

30. A compound is 17.2% C, 1.44% H, and 81.4% F. Find its empirical formula.

Answer: [ ]

Section 7.9: Determination of a Molecular Formula

31. A coolant widely used in automobile engines is 38.7% carbon, 9.7% hydrogen, and 51.6% oxygen. Its molar mass is 62.0 g/mol. What is the molecular formula of the compound?

Answer: [ ]

Name: ______________________ Date: ______________________

ID: ______________________ Section: ______________________

32. A compound is 73.1% chlorine, 24.8% carbon, and the balance is hydrogen. If the molar mass of the compound is 97 g/mol, find the molecular formula.

Answer: [ ]

General Questions

33. The stable form of elemental phosphorus is a tetratomic molecule. Calculate the number of molecules and atoms in 85.0 g P_4.

Answer: [ ]

Name: ______________________ Date: ______________________

ID: ______________________ Section: ______________________

34. Is it reasonable to set a dinner table with one mole of salt, NaCl, in a salt shaker with a capacity of 2 oz? How about one mole of sugar, $C_{12}H_{22}O_{11}$, in a sugar bowl with a capacity of 10 oz?

More Challenging Problems

35. Assuming gasoline to be pure octane, C_8H_{18} (actually, it is a mixture of many substances), an automobile getting 25.0 miles per gallon would consume 5.62×10^{23} molecules per mile. Calculate the mass of this amount of fuel.

Answer: ______________________

Name: ______________________ Date: ______________________

ID: ______________________ Section: ______________________

36. $Co_aS_bO_c \cdot X\,H_2O$ is the general formula of a certain hydrate. When 43.0 g of the compound is heated to drive off the water, 26.1 g of anhydrous compound is left. Further analysis shows that the percentage composition of the anhydrate is 42.4% Co, 23.0% S, and 34.6% O. Find the empirical formula of (a) the anhydrous compound and (b) the hydrate. (*Hint:* Treat the anhydrous compound and water just as you have treated elements in calculating X in the formula of the hydrate.)

Answer (a): ______________________

Answer (b): ______________________

Chapter 8

Reactions and Equations

☑ TARGET CHECK 8.1-WB

Is the following true or false? Explain your reasoning.
The equation $H_2 + O_2 \rightarrow H_2O$ may be balanced by changing it to $H_2 + O_2 \rightarrow H_2O_2$.

EXAMPLE 8.1-WB

Balance the equation

_____ $BiOCl(aq)$ + _____ $H_2S(g)$ → _____ $Bi_2S_3(s)$ + _____ $H_2O(\ell)$ + _____ $HCl(aq)$

Take it all the way without suggestions this time.

$2\ BiOCl(aq) + 3\ H_2S(g) \rightarrow Bi_2S_3(s) + 2\ H_2O(\ell) + 2\ HCl(aq)$

Step 1: Put a "1" before the starting formula.
$BiOCl(aq) + H_2S(g) \rightarrow 1\ Bi_2S_3(s) + H_2O(\ell) + HCl(aq)$

Step 2: Balance elements in compounds.
Elements in Bi_2S_3 that are in only one other compound (Bi and S).
$2\ BiOCl(aq) + 3\ H_2S(g) \rightarrow 1\ Bi_2S_3(s) + H_2O(\ell) + HCl(aq)$

Other elements in one compound formula on each side (O and Cl).
$2\ BiOCl(aq) + 3\ H_2S(g) \rightarrow 1\ Bi_2S_3(s) + 2\ H_2O(\ell) + 2\ HCl(aq)$

Step 3: Balance uncombined elements. (None)

Step 4: Clear fractions; remove 1s.
$2\ BiOCl(aq) + 3\ H_2S(g) \rightarrow Bi_2S_3(s) + 2\ H_2O(\ell) + 2\ HCl(aq)$

Step 5: Check.
2 Bi, 2 O, 2 Cl, 6 H, 3 S on both sides. ✓

Note that, again, the last element is balanced by the time we get to it.

EXAMPLE 8.2-WB

Plants convert water and carbon dioxide into sugar and oxygen in a process known as photosynthesis. This reaction is the primary mechanism by which the energy of the sun is incorporated into chemical bonds in substances on earth. A representative photosynthesis reaction is:

$$\_\_\_\_ \ H_2O(\ell) + \_\_\_\_ \ CO_2(g) \rightarrow \_\_\_\_ \ C_6H_{12}O_6(s) + \_\_\_\_ \ O_2(g)$$

Balance the equation.

Add the appropriate coefficients.

$$6\ H_2O(\ell) + 6\ CO_2(g) \rightarrow C_6H_{12}O_6(s) + 6\ O_2(g)$$

"1" before starting formula:	$H_2O(\ell) + CO_2(g) \rightarrow 1\ C_6H_{12}O_6(s) + O_2(g)$
Balance C:	$H_2O(\ell) + 6\ CO_2(g) \rightarrow 1\ C_6H_{12}O_6(s) + O_2(g)$
Balance H:	$6\ H_2O(\ell) + 6\ CO_2(g) \rightarrow 1\ C_6H_{12}O_6(s) + O_2(g)$
Balance O:	$6\ H_2O(\ell) + 6\ CO_2(g) \rightarrow 1\ C_6H_{12}O_6(s) + 6\ O_2(g)$
Clear fractions; remove 1's:	$6\ H_2O(\ell) + 6\ CO_2(g) \rightarrow C_6H_{12}O_6(s) + 6\ O_2(g)$
Final check:	12 H, 18 O, 6 C on each side

Note that, in the "Balance O" step, the 18 O on the left $(6 + 6 \times 2)$ require 18 O on the right. Six O already are present in the 1 $C_6H_{12}O_6$ molecule. The remaining 12 needed are divided by 2 to arrive at a coefficient of 6 for O_2.

☑ TARGET CHECK 8.2-WB

Consider the chemical equation for the reaction of carbon and oxygen: $C(s) + O_2(g) \rightarrow CO_2(g)$
Write a word description of this equation on the (a) particulate, (b) molar, and (c) macroscopic levels.

EXAMPLE 8.3-WB

Hydrogen chloride gas is formed by direct combination of its gaseous elements. Write the equation.

Again, recognize that this is a combination reaction. Two elements combine to form one product compound. Write the formulas, and then balance the equation.

$$H_2(g) + Cl_2(g) \rightarrow 2\ HCl(g)$$

EXAMPLE 8.4-WB

Solid magnesium hydroxide is formed when solid magnesium oxide combines with water. Write the equation.

Start with the unbalanced equation.

$MgO(s)$ + $H_2O(\ell)$ → $Mg(OH)_2(s)$

Now quantify the equation by adding the appropriate coefficients.

$MgO(s) + H_2O(\ell) \rightarrow Mg(OH)_2(s)$

Once again, the qualitative description matches the quantitative description because all coefficients are 1. This example also illustrates how compounds can react with each other in combination reactions, too.

EXAMPLE 8.5-WB

A common laboratory procedure for producing oxygen is heating solid potassium chlorate. Solid potassium chloride is left behind. What is the equation for this reaction?

Read the description of the reaction carefully to be sure you identify the reactants and products correctly, and then write their formulas where they should be.

$KClO_3(s)$ → $O_2(g)$ + $KCl(s)$

Complete the equation.

$2\ KClO_3(s) \rightarrow 3\ O_2(g) + 2\ KCl(s)$

Notice that the original equation has 1 K and 1 Cl on each side; they are already balanced. The only element that remains is oxygen, and it is in the elemental form on the right. The O_2 coefficient that balances oxygen balances the equation. To balance the 3 O's in one $KClO_3$ with O_2 units tales 1 1/2 units, or 3/2 O_2: $KClO_3(s) \rightarrow 3/2\ O_2(g) + KCl(s)$. The fractional coefficient disappears when the entire equation is multiplied by 2.

EXAMPLE 8.6-WB

When heated, solid magnesium carbonate reacts to form solid magnesium oxide and carbon dioxide gas. Write the equation for this reaction.

Can you recognize from the reaction description that this is a decomposition reaction? Write and balance the equation.

$MgCO_3(s) \rightarrow MgO(s) + CO_2(g)$

EXAMPLE 8.7-WB

Write the equation for the complete oxidation of butanol, $C_4H_9OH(\ell)$.

This equation is like the equation in textbook Example 8.2—actually, a little easier. Write the balanced equation.

$$C_4H_9OH(\ell) + 6\ O_2(g) \rightarrow 4\ CO_2(g) + 5\ H_2O(\ell)$$

If you did not get this equation, you didn't check your work or you would have caught your error. You may have counted only nine hydrogen atoms in C_4H_9OH (there are ten), or you may have overlooked the oxygen in C_4H_9OH in finding the coefficient of O_2.

EXAMPLE 8.8-WB

Solid zinc metal is placed in aqueous hydrochloric acid. Write the equation for the reaction that occurs.

Once you've written the formulas of the reactants, you will be able to recognize that this is a single-replacement reaction. Start there.

$$Zn(s) + HCl(aq) \rightarrow$$

The reactants are in the form A + BX, where A is an element and BX is an ionic compound.

The reactant that is an element, Zn(s), will become part of a compound in the products. One of the elements in the compound HCl(aq) will become an uncombined product element. Will zinc replace the H^+ ion or the Cl^- ion in HCl? Why?

Zinc will replace the H^+ ion in HCl because the zinc ion, Zn^{2+}, is a positive ion so it must replace the positive ion in the compound.

Now write the formulas of the products after the reaction arrow above.

$$Zn(s) + \quad HCl(aq) \rightarrow \quad ZnCl_2(aq) + \quad H_2(g)$$

The zinc ion, Zn^{2+}, must be combined with two chloride ions, Cl^-, to produce a neutral compound. The other product, the element hydrogen, H_2, occurs in nature as a diatomic molecule. We followed the assumption given in textbook Example 8.8 in writing the aqueous state for $ZnCl_2(aq)$.

Complete the equation above by balancing.

$$Zn(s) + 2\ HCl(aq) \rightarrow ZnCl_2(aq) + H_2(g)$$

EXAMPLE 8.9-WB

Write the equation for the reaction between chlorine gas and a solution of sodium bromide.

The formulas of the reactants, please . . .

$Cl_2(g)$ + NaBr(aq) →

This time the elemental reactant is a nonmetal, so it will replace the nonmetal in the compound. Write the formulas of the products, both of which are in aqueous solution.

$Cl_2(g)$ + NaBr(aq) → NaCl(aq) + $Br_2(aq)$

Finish by balancing the equation.

Cl_2 (g) + 2 NaBr(aq) → 2 NaCl(aq) + $Br_2(aq)$

EXAMPLE 8.10-WB

A precipitate forms when solutions of potassium hydroxide and aluminum nitrate are combined. Write the equation.

Do not be concerned about the states of the reactants and products. We will show them in the answer. Start with the formulas of the reactants.

KOH(aq) + $Al(NO_3)_3(aq)$ →

Now do the double replacement. In other words, swap partners. Pair one positive ion with the opposite negative ion and pair the other cation with a different anion. Be sure to put the appropriate numbers of each ion in each product compound to make them electrically neutral.

KOH(aq) + $Al(NO_3)_3(aq)$ → $KNO_3(aq)$ + $Al(OH)_3(s)$

Aluminum hydroxide is the solid precipitate in this reaction.

Complete the example by balancing the equation and checking the balance.

---

3 KOH(aq) + $Al(NO_3)_3(aq)$ → 3 $KNO_3(aq)$ + $Al(OH)_3(s)$

Three nitrate ions in $Al(NO_3)_3$ require a coefficient of 3 for KNO_3. This, in turn, creates a need for a coefficient of 3 for KOH. The three hydroxide ions on each side balance, as does the one aluminum ion on each side. The balance checks.

EXAMPLE 8.11-WB

Write the equation for the reaction between a solution of barium hydroxide and a solution of nitric acid.

As always, start with the formulas of the reactants.

$Ba(OH)_2(aq)$ + $HNO_3(aq)$ →

After writing the formulas of the reactants, you can see that they are in the general form cation-anion + cation-anion, which indicates a double-replacement reaction. On further inspection, one of the cations is H^+ and one anion is OH^-, so this will be a neutralization reaction, forming HOH or, in proper form for a molecular compound, H_2O. Switch partners and write the products.

$Ba(OH)_2(aq)$ + $HNO_3(aq)$ → $Ba(NO_3)_2(aq)$ + $HOH(\ell)$

We've written the formula of water as $HOH(\ell)$ as a temporary substitute for the correct formula to make balancing a bit easier. We must keep in mind that we need to change the formula back to $H_2O(\ell)$ by the end of the balancing process.

Balance the equation. You can balance the hydroxide and nitrate ions as complete units.

$Ba(OH)_2(aq) + 2\ HNO_3(aq) \rightarrow Ba(NO_3)_2(aq) + 2\ HOH(\ell)$

Check the balance. If you used our $HOH(\ell)$ balancing technique, change the formula into its proper form after the check.

---

1 Ba, 2 OH, 2 H, and 2 NO_3 on each side. The equation is balanced.
$Ba(OH)_2(aq) + 2\ HNO_3(aq) \rightarrow Ba(NO_3)_2(aq) + 2\ H_2O(\ell)$

Name: ____________________ Date: ____________________

ID: ____________________ Section: ____________________

Questions, Exercises, and Problems

Section 8.3: Interpreting Chemical Equations

1. Give a description, in words, of the following chemical equation on both the particulate and molar levels: $2\ O_3 \rightarrow 3\ O_2$. (The formula of ozone is O_3.)

Questions 2 to 14: Write the equation for each reaction described.

Section 8.5: Combination Reactions

2. Lithium combines with oxygen to form lithium oxide.

3. Boron combines with oxygen to form diboron trioxide.

4. Calcium combines with bromine to make calcium bromide.

Section 8.6: Decomposition Reactions

5. Pure hydrogen iodide decomposes spontaneously to its elements.

6. Barium peroxide, BaO_2, breaks down into barium oxide and oxygen.

Name: ______________________ Date: ______________________

ID: ______________________ Section: ______________________

Section 8.7: Complete Oxidation or Burning of Organic Compounds

7. Propane, C_3H_8, a component of "bottled gas," is burned as a fuel in heating homes.

8. Ethanol, C_2H_5OH, the alcohol in alcoholic beverages, is oxidized.

Section 8.8: Single-Replacement Oxidation–Reduction (Redox) Reactions

9. Calcium reacts with hydrobromic acid.

10. Chlorine gas is bubbled through an aqueous solution of potassium iodide.

Section 8.9: Double-Replacement Precipitation Reactions

It is not necessary for you to identify the precipitates formed in Questions 11 and 12.

11. Calcium chloride and potassium fluoride solutions react to form a precipitate.

12. Milk of magnesia is the precipitate that results when sodium hydroxide and magnesium bromide solutions are combined.

Section 8.10: Double-Replacement Neutralization Reactions

13. Sulfuric acid reacts with barium hydroxide solution.

Name: ______________________ Date: ______________________

ID: ______________________ Section: ______________________

14. Sodium hydroxide is added to phosphoric acid.

Unclassified Reactions

Questions 15 to 32: Write the equation for the reaction described or for the most likely reaction between given reactants.

15. Lead(II) nitrate solution reacts with a solution of sodium iodide.

16. The fuel butane, C_4H_{10}, burns.

17. Sulfurous acid decomposes spontaneously to sulfur dioxide and water.

18. Potassium reacts violently with water. (*Hint:* Think of water as an acid with the formula HOH.)

19. Zinc metal is placed in a silver chlorate solution.

20. Ammonium sulfide is added to a solution of copper(II) nitrate.

21. Phosphorus tribromide is produced when phosphorus reacts with bromine.

22. When calcium hydroxide (sometimes called slaked lime) is heated, it forms calcium oxide—lime—and water vapor.

Name: ______________________ Date: ______________________

ID: ______________________ Section: ______________________

23. Glycerine, $C_3H_8O_3$—used in making soap, cosmetics, and explosives—is completely oxidized.

24. Powdered antimony (Z = 51) ignites when sprinkled into chlorine gas, producing antimony(III) chloride.

25. A solution of potassium hydroxide reacts with a solution of zinc chloride.

26. Aluminum carbide, Al_4C_3, is the product of the reaction of its elements.

27. A solution of lithium sulfite is mixed with a solution of sodium phosphate.

28. A solution of chromium(III) nitrate is one of the products of the reaction between metallic chromium and aqueous tin(II) nitrate.

29. Sulfuric acid is produced when sulfur trioxide reacts with water.

The remaining reactions are not readily placed into one of the six classifications used in this lesson. Nevertheless, enough information is given for you to write the equations.

30. A solid oxide of iron, Fe_3O_4, and hydrogen are the products of the reaction between iron and steam.

31. Aluminum carbide, Al_4C_3, reacts with water to form aluminum hydroxide and methane, CH_4.

Name: ______________________ Date: ______________________

ID: ______________________ Section: ______________________

32. Magnesium nitride and hydrogen are the products of the reaction between magnesium and ammonia.

General Questions

33. Each reactant or pair of reactants listed below is *potentially* able to participate in one type of reaction described in this lesson. In each case, name the type of reaction and complete the equation. In part e, a compound containing Cu^{2+} is formed from the reaction of the corresponding metal.

a) NH_4Cl + $AgNO_3$ →

b) Ca + Cl_2 →

c) F_2 + NaI →

d) $Zn(NO_3)_2$ + $Ba(OH)_2$ →

e) Cu + $NiCl_2$ →

34. Acid rain is rainfall that contains sulfuric acid originating from organic fuels that contain sulfur. The process occurs in three major steps. The sulfur is first burned, forming sulfur dioxide. In sunlight, the sulfur dioxide reacts with oxygen in the air to produce sulfur trioxide. When rainwater falls through the sulfur trioxide, the reaction produces sulfuric acid. Write the equation for each step in the process and tell what kind of reaction it is.

Name: ______________________ Date: ______________________

ID: ______________________ Section: ______________________

35. The tarnish that appears on silver is silver sulfide, which is formed when silver is exposed to sulfur-bearing compounds in the presence of oxygen in the air. When hydrogen sulfide is the sulfur-bearing reacting compound, water is the second product. Write the equation for the reaction.

36. One source of the pure tungsten (Z = 74) filament used in light bulbs is tungsten(VI) oxide. It is heated with hydrogen at high temperatures. The hydrogen reacts with the oxygen in the oxide, forming steam. Write the equation for the reaction and classify it as one of the reaction types discussed in this chapter.

More Challenging Problems

Questions 37 and 38: Write the equation for each reaction described.

37. Write the equation for the neutralization reaction in which barium nitrate is the salt formed.

38. Only the first hydrogen comes off in the reaction between sulfamic acid, HNH_2SO_3, and potassium hydroxide. (Figure out the formula of the anion from the acid and use it to write the formula of the salt formed.)

Answers to Target Checks

1. False. Never change a chemical formula to balance an equation. H_2O_2 is a real compound, but it has nothing to do with this reaction.
2. (a) Particulate: One carbon atom combines with one oxygen molecule to form one carbon dioxide molecule. (b) Molar: One mole of carbon atoms combines with one mole of oxygen molecules to form one mole of carbon dioxide molecules. (c) Macroscopic: Twelve grams of carbon combines with 32 grams of oxygen to form 44 grams of carbon dioxide. (We used the mass of one mole of each reactant and product; many other answers are possible.)

Chapter 9

Quantity Relationships in Chemical Reactions

EXAMPLE 9.1-WB

Ammonia is formed directly from its elements. How many moles of hydrogen are needed to produce 4.20 moles of ammonia?

The procedure is exactly the same as in textbook Example 9.1. Balance the equation and complete the problem.

$N_2 + 3\ H_2 \rightarrow 2\ NH_3$

GIVEN: 4.20 mol NH_3 *WANTED*: mol H_2

PER/PATH: mol NH_3 $\xrightarrow{3 \text{ mol } H_2/2 \text{ mol } NH_3}$ mol H_2

$$4.20 \text{ mol } NH_3 \times \frac{3 \text{ mol } H_2}{2 \text{ mol } NH_3} = 6.30 \text{ mol } H_2$$

EXAMPLE 9.3-WB

How many moles of H_2O will be produced in a heptane-burning reaction that also yields 115 grams of CO_2?

Begin with the starting steps.

$C_7H_{16}(\ell) + 11\ O_2(g) \rightarrow 7\ CO_2(g) + 8\ H_2O(\ell)$

GIVEN: 115 g CO_2 *WANTED:* mol H_2O

PER/PATH: g CO_2 $\xrightarrow{44.01 \text{ g } CO_2/\text{mol } CO_2}$ mol CO_2 $\xrightarrow{8 \text{ mol } H_2O/7 \text{ mol } CO_2}$ mol H_2O

This time the wanted quantity is moles of H_2O, which is reached by using only *Step 1* and *Step 2* of the stoichiometry path.

Set up the first conversion, g $CO_2 \rightarrow$ mol CO_2, but do not calculate the answer.

$$115 \text{ g } CO_2 \times \frac{1 \text{ mol } CO_2}{44.01 \text{ g } CO_2} \times \underline{\hspace{6cm}} =$$

Now you can complete the problem by changing moles of CO_2 to moles of H_2O.

$$115 \text{ g } CO_2 \times \frac{1 \text{ mol } CO_2}{44.01 \text{ g } CO_2} \times \frac{8 \text{ mol } H_2O}{7 \text{ mol } CO_2} = 2.99 \text{ mol } H_2O$$

EXAMPLE 9.4-WB

How many grams of calcium fluoride will precipitate from a combination of two solutions, one containing 27.9 grams of sodium fluoride and the other an excess of calcium nitrate?

Begin with writing and balancing the equation for the reaction.

$$2 \text{ NaF(aq)} + \text{Ca(NO}_3)_2\text{(aq)} \rightarrow 2 \text{ NaNO}_3\text{(aq)} + \text{CaF}_2\text{(s)}$$

Now identify what you know and what you are looking for. Write the *GIVEN* and *WANTED.*

GIVEN: 27.9 g NaF *WANTED:* g CaF_2

Your next task is to write the *PER/PATH* that will take you from the given 27.9 g NaF to the wanted g CaF_2. You'll have to calculate a couple of molar masses to get all of the *PER* expressions.

PER/PATH: g NaF $\xrightarrow{41.99\text{ g NaF/mol NaF}}$ mol NaF $\xrightarrow{1\text{ mol CaF}_2/2\text{ mol NaF}}$ mol CaF_2 $\xrightarrow{78.08\text{ g CaF}_2/\text{mol CaF}_2}$ g CaF_2

You've completed the bulk of the work. Now it's just a matter of executing your *PLAN.* Set up the calculation, following your *PER/PATH,* and then calculate your final answer. Don't forget to think about significant figures.

$$27.9\text{ g NaF} \times \frac{1\text{ mol NaF}}{41.99\text{ g NaF}} \times \frac{1\text{ mol CaF}_2}{2\text{ mol NaF}} \times \frac{78.08\text{ g CaF}_2}{\text{mol CaF}_2} = 25.9\text{ g CaF}_2$$

EXAMPLE 9.5-WB

Consider the complete combustion of liquid cyclopentane, $C_5H_{10}(\ell)$. If the reaction consumes 4.12 kilograms of oxygen, how many kilograms of water are produced?

Recall the procedure for writing combustion reactions (review Section 8.7, if necessary) and write a balanced equation.

$$2\ C_5H_{10}(\ell) + 15\ O_2(g) \rightarrow 10\ CO_2(g) + 10\ H_2O(\ell)$$

Complete the *PLAN* for solving this problem. As in textbook Example 9.5, do the same thing that you would do with grams and moles, but change the units to kilograms and kilomoles.

GIVEN: 4.12 kg O_2 *WANTED:* kg H_2O

PER/PATH: kg O_2 $\xrightarrow{32.00\text{ kg O}_2/\text{kmol O}_2}$ kmol O_2 $\xrightarrow{10\text{ kmol H}_2\text{O}/10\text{ kmol CO}_2}$ kmol CO_2 $\xrightarrow{18.02\text{ kg H}_2\text{O}/\text{kmol H}_2\text{O}}$ kg H_2O

Finish the problem.

$$4.12 \text{ kg } O_2 \times \frac{1 \text{ kmol } O_2}{32.00 \text{ kg } O_2} \times \frac{10 \text{ kmol } H_2O}{10 \text{ kmol } O_2} \times \frac{18.02 \text{ kg } H_2O}{\text{kmol } H_2O} = 2.32 \text{ kg } H_2O$$

EXAMPLE 9.6-WB

Calculate the number of moles of argon in 61.9 liters, measured at STP.

Write the *PLAN* for your solution.

GIVEN: 61.9 L *WANTED:* mol *PER/PATH:* $\text{L} \xrightarrow{22.4 \text{ L/mol}} \text{mol}$

Calculate the answer.

$$61.9 \text{ L} \times \frac{1 \text{ mol}}{22.4 \text{ L}} = 2.76 \text{ mol}$$

As in textbook Example 9.6, the identity of the gas has no bearing on the answer to the question.

EXAMPLE 9.7-WB

How many grams of sodium hydroxide can be recovered as a by-product of the chemical change that produces 71.9 liters of chlorine, measured at STP, by the reaction $2\text{ NaCl(aq)} + 2\text{ H}_2\text{O}(\ell) \rightarrow 2\text{ NaOH(aq)} + \text{Cl}_2\text{(g)} + \text{H}_2\text{(g)}$?

Start by listing the *GIVEN* and *WANTED* for this problem.

GIVEN: 71.9 L Cl_2 *WANTED:* g NaOH

Now write out a *PER/PATH* that takes you from liters of chlorine to grams of sodium hydroxide.

PER/PATH: L Cl_2 $\xrightarrow{22.4 \text{ L } Cl_2/\text{mol } Cl_2}$ mol Cl_2 $\xrightarrow{2 \text{ mol NaOH}/1 \text{ mol } Cl_2}$ mol NaOH $\xrightarrow{40.00 \text{ g NaOH/mol NaOH}}$ g NaOH

Complete the problem with the setup and final answer.

$$71.9 \text{ L } Cl_2 \times \frac{1 \text{ mol } Cl_2}{22.4 \text{ L } Cl_2} \times \frac{2 \text{ mol NaOH}}{1 \text{ mol } Cl_2} \times \frac{40.00 \text{ g NaOH}}{\text{mol NaOH}} = 257 \text{ g NaOH}$$

EXAMPLE 9.8-WB

Phosphine, PH_3, is an extremely poisonous gas that decomposes by the reaction $4\ PH_3(g) \rightarrow P_4(g) + 6\ H_2(g)$. Calculate the volume of phosphine that must decompose to produce 11.3 liters of hydrogen if both volumes are measured at STP.

List the *GIVEN* and *WANTED.*

GIVEN: 11.3 L H_2 *WANTED:* volume PH_3 (assume L)

If you were given some number of *moles* of H_2 and you were asked for *moles* of PH_3, what would you write as the *PER* expression for the *PATH* mol $H_2 \rightarrow$ mol PH_3?

---

4 mol PH_3/6 mol H_2 (or the inverse, 6 mol H_2/4 mol PH_3)

The mole ratio comes from the balanced chemical equation.

Since V ∝ n at constant temperature and pressure, the mole ratio that you just wrote is also a volume ratio. Write the *PER/PATH* to go from your *GIVEN* to your *WANTED.*

$$L\ H_2 \xrightarrow{4\ L\ PH_3/6\ L\ H_2} L\ PH_3$$

The coefficients from the balanced chemical equation can be interpreted as a volume ratio as long at the reaction involves gases measured at the same temperature and pressure.

Complete the problem with the setup and the answer.

$$11.3\ L\ H_2 \times \frac{4\ L\ PH_3}{6\ L\ H_2} = 7.53\ L\ PH_3$$

EXAMPLE 9.9-WB

Calculate the volume of carbon dioxide at 169°C and 744 torr released by decomposing 155 g magnesium carbonate by the decomposition reaction of magnesium carbonate to form magnesium oxide and carbon dioxide.

List the *GIVEN* and *WANTED* for this question.

GIVEN: 155 g $MgCO_3$ *WANTED:* volume CO_2 (assume L)

Notice what type of problem this is. First, it's a stoichiometry problem because the question asks about quantities of different species in a chemical change. Second, the given is a solid with its quantity stated in grams, and the wanted is a gas, measured as volume at a specified temperature and pressure. The stoichiometry part of the problem requires you to write and balance the chemical equation for the reaction. Do so now.

$$MgCO_3(s) \rightarrow MgO(s) + CO_2(g)$$

Starting from the given mass of solid, the initial steps in the *PER/PATH* are the conversion to moles of the given and then to moles of the wanted. Complete those steps, including the necessary *PER* expressions.

$$\text{g } MgCO_3 \xrightarrow{84.32 \text{ g } MgCO_3/\text{mol } MgCO_3} \text{mol } MgCO_3$$

$$\xrightarrow{1 \text{ mol } CO_2/1 \text{ mol } MgCO_3} \text{mol } CO_2 \xrightarrow{\hspace{4cm}}$$

At this point in your setup, you have moles of CO_2, a species that is in the gaseous state at the conditions given in the problem. You want the volume of this gas. What do you know that links the volume of a gas and the number of moles of that gas?

22.4 L/mol of a gas at STP

Extend your *PER/PATH* to include this *PER* expression, using the arrow provided above.

$$\text{g } MgCO_3 \xrightarrow{84.32 \text{ g } MgCO_3/\text{mol } MgCO_3} \text{mol } MgCO_3$$

$$\xrightarrow{1 \text{ mol } CO_2/1 \text{ mol } MgCO_3} \text{mol } CO_2 \xrightarrow{22.4 \text{ L } CO_2/\text{mol } CO_2} \text{L } CO_2 \text{ (STP)}$$

The next phase of solving this problem will involve a combined gas laws calculation that's solved by algebra, so it's a good idea to complete the calculation for the dimensional analysis part of the solution now. Find the number of liters of carbon dioxide produced at STP.

$$155 \text{ g } MgCO_3 \times \frac{1 \text{ mol } MgCO_3}{84.32 \text{ g } MgCO_3} \times \frac{1 \text{ mol } CO_2}{1 \text{ mol } MgCO_3} \times \frac{22.4 \text{ L } CO_2}{\text{mol } CO_2} = 41.2 \text{ L } CO_2 \text{ (STP)}$$

Now you have the STP volume of CO_2, and you're asked for the 169°C and 744 torr volume of CO_2. This is a combined gas laws equation problem. Set up a table of initial and final values and take it to completion.

	Volume	Temperature	Pressure
Initial Value (1)	41.2 L	0°C; 273 K	1 atm (exact)
Final Value (2)	V_2	169°C; 442 K	744 torr

$$V_2 = \frac{P_1V_1T_2}{P_2T_1} = V_1 \times \frac{P_1}{P_2} \times \frac{T_2}{T_1} = 41.2\text{ L} \times \frac{760\text{ torr}}{744\text{ torr}} \times \frac{442\text{ K}}{273\text{ K}} = 68.1\text{ L}$$

EXAMPLE 9.10-WB

The combination reaction of nitrogen monoxide and chlorine gases produces NOCl(g), commonly known as nitrosyl chloride. What maximum volume of nitrosyl chloride, measured at 35°C and 0.943 atm, can be obtained from the reaction of 5.17 L of chlorine gas at 18°C and 49.0 atm?

Start by writing and balancing the equation for the reaction described, and then set up a table of initial and final values for volume, temperature, and pressure of chlorine to adjust its starting conditions to those at which the product will be measured.

$$2\text{ NO(g)} + \text{Cl}_2\text{(g)} \rightarrow 2\text{ NOCl(g)}$$

	Volume	Temperature	Pressure
Initial Value (1)	5.17 L	18°C; 291 K	49.0 atm
Final Value (2)	V_2	35°C; 308 K	0.943 atm

Calculate the volume that chlorine would occupy if it were measured at 35°C and 0.943 atm.

$$V_2 = V_1 \times \frac{P_1}{P_2} \times \frac{T_2}{T_1} = 5.17 \text{ L} \times \frac{49.0 \text{ atm}}{0.943 \text{ atm}} \times \frac{308 \text{ K}}{291 \text{ K}} = 284 \text{ L}$$

Now prepare to complete the solution of the problem. *PLAN* the remainder of the work needed to get the answer.

GIVEN: 284 L Cl_2 *WANTED:* volume NOCl (assume L)

PER/PATH: L Cl_2 $\xrightarrow{2 \text{ L NOCl}/1 \text{ L } Cl_2}$ L NOCl

Complete the example by calculating the final answer.

$$284 \text{ L } Cl_2 \times \frac{2 \text{ L NOCl}}{1 \text{ L } Cl_2} = 568 \text{ L NOCl}$$

EXAMPLE 9.11-WB

Calculate the theoretical yield and the percentage yield of nitrogen oxide if 49.2 grams of nitrogen dioxide yield 8.90 grams of nitrogen oxide in the reaction of nitrogen dioxide and water to form nitric acid and nitrogen oxide.

Determining the theoretical yield is the first step. Complete the *PLAN*.

$3\ NO_2 + H_2O \rightarrow 2\ HNO_3 + NO$

GIVEN: 49.2 g NO_2 *WANTED:* g NO

PER/PATH: $\text{g } NO_2 \xrightarrow{46.01 \text{ g } NO_2/\text{mol } NO_2} \text{mol } NO_2 \xrightarrow{1 \text{ mol NO}/3 \text{ mol } NO_2} \text{mol NO} \xrightarrow{30.01 \text{ g NO/mol NO}} \text{g NO}$

Now calculate the theoretical yield.

$$49.2 \text{ g } NO_2 \times \frac{1 \text{ mol } NO_2}{46.01 \text{ g } NO_2} \times \frac{1 \text{ mol NO}}{3 \text{ mol } NO_2} \times \frac{30.01 \text{ g NO}}{\text{mol NO}} = 10.7 \text{ g NO theo}$$

Finish the problem with a *PLAN* for and calculation of the percentage yield.

GIVEN: 8.90 g NO act; 10.7 g NO theo *WANTED:* % yield

EQUATION: $\% \text{ yield} = \frac{\text{g act}}{\text{g theo}} \times 100 = \frac{8.90 \text{ g}}{10.7 \text{ g}} \times 100 = 83.2\%$

EXAMPLE 9.13-WB

Sodium nitrite is produced from sodium nitrate by a decomposition reaction. Solid sodium nitrate decomposes to solid sodium nitrite and oxygen gas. How many grams of sodium nitrate must be used to produce an actual yield of 60.0 g sodium nitrite if the percent yield is 76.3%?

Begin by writing the *GIVEN* and *WANTED.* Also write the equation for the reaction.

GIVEN: 60.0 g $NaNO_2$ (act) *WANTED:* g $NaNO_3$

$2\ NaNO_3(s) \rightarrow 2\ NaNO_2(s) + O_2(g)$

Your *PER/PATH* will start with a conversion between g $NaNO_2$ (act) and g $NaNO_2$ (theo), using the percent yield, and then it will follow the standard stoichiometry three-step procedure. Finish the *PLAN* for this problem.

PER/PATH: g $NaNO_2$ (act) $\xrightarrow{76.3 \text{ g } NaNO_2 \text{ (act)}/100 \text{ g } NaNO_2 \text{ (theo)}}$ g $NaNO_2$ (theo) $\xrightarrow{69.00 \text{ g } NaNO_2/\text{mol } NaNO_2}$ mol $NaNO_2$ $\xrightarrow{2 \text{ mol } NaNO_3/2 \text{ mol } NaNO_2}$ mol $NaNO_3$ $\xrightarrow{85.00 \text{ g } NaNO_3/\text{mol } NaNO_3}$ g $NaNO_3$

You are now ready to set up and solve the problem.

$$60.0 \text{ g } NaNO_2 \text{ (act)} \times \frac{100 \text{ g } NaNO_2 \text{ (theo)}}{76.3 \text{ g } NaNO_2 \text{ (act)}} \times \frac{1 \text{ mol } NaNO_2}{69.00 \text{ g } NaNO_2} \times \frac{2 \text{ mol } NaNO_3}{2 \text{ mol } NaNO_2} \times \frac{85.00 \text{ g } NaNO_3}{\text{mol } NaNO_3} = 96.9 \text{ g } NaNO_3$$

☑ TARGET CHECK 9.1-WB

(a) If three moles of C are combined with four moles of D and allowed to react according to the equation $C + D \rightarrow CD$, how many moles of CD will result? How many moles of which reactant will remain? (b) Three moles of E and four moles of F are placed in a reaction vessel. They react according to the equation $E + 2F \rightarrow EF_2$. How many moles of product result? How many moles of which reactant are left? (c) In parts (a) and (b) three moles of one reactant react with four moles of another reactant. The number of moles of product molecules is different. Explain why.

EXAMPLE 9.15-WB

A student drops 0.160 mole of calcium metal into a solution that contains 0.246 mole of HCl. Assume the reaction proceeds until the limiting reactant is consumed. (a) How many moles of hydrogen gas will be produced? (b) Identify the reactant that is in excess and calculate the number of moles of that reactant that will remain unreacted.

Write and balance the reaction equation.

$Ca + 2\ HCl \rightarrow CaCl_2 + H_2$

Fill in the first line below. You can disregard the $CaCl_2$ column because the problem does not ask about it.

	Ca	+	$2\ HCl$	$\rightarrow$	H_2	$(+\ CaCl_2)$
Moles at start						
Moles used (–), produced (+)						
Moles at end						

	Ca	+	$2\ HCl$	$\rightarrow$	H_2	$(+\ CaCl_2)$
Moles at start	0.160		0.246		0	
Moles used (–), produced (+)						
Moles at end						

The next step is to identify the limiting reactant. You can either start with moles of Ca and find the number of moles of HCl needed for a complete reaction or start with moles of HCl and determine the number of moles of Ca required. Do one of these calculations (we'll show solutions for both), state which reactant is limiting, and then fill in the center line in the table above.

$$0.160 \text{ mol Ca} \times \frac{2 \text{ mol HCl}}{1 \text{ mol Ca}} = 0.320 \text{ mol HCl}$$

A total of 0.320 mol HCl is needed to complete react with 0.160 mol Ca, and only 0.246 mol HCl is available, so HCl is the limiting reactant.

$$0.246 \text{ mol HCl} \times \frac{1 \text{ mol Ca}}{2 \text{ mol HCl}} = 0.123 \text{ mol Ca}$$

Only 0.123 mol of the available 0.160 mol of Ca will react with the initial 0.246 mole of HCl, so HCl is the limiting reactant.

	Ca	+	2 HCl	→	H_2	($+ CaCl_2$)
Moles at start	0.160		0.246		0	
Moles used (–), produced (+)	– 0.123		– 0.246		+ 0.123	
Moles at end						

Complete the preceding table and give your final answer to both parts of the problem.

	Ca	+	2 HCl	→	H_2	($+ CaCl_2$)
Moles at start	0.160		0.246		0	
Moles used (–), produced (+)	– 0.123		– 0.246		+ 0.123	
Moles at end	0.037		0		0.123	

(a) 0.123 mol H_2 is produced. (b) Ca is in excess, and 0.037 mol remains unreacted.

EXAMPLE 9.16-WB

A solution that contains 29.0 g of calcium nitrate is added to a solution that contains 33.0 g of sodium fluoride. Calculate the number of grams of calcium fluoride that will precipitate. Which reactant is in excess? How many grams of excess reactant are left over?

Write and balance the equation for this double-replacement precipitation reaction.

$$Ca(NO_3)_2(aq) + 2\ NaF(aq) \rightarrow CaF_2(s) + 2\ NaNO_3(aq)$$

Begin a table with columns for each reactant and the solid product. Start by listing grams at start, molar mass, and moles at start for each. Leave enough space in your table to add three more lines.

	$Ca(NO_3)_2(aq)$ +	2 $NaF(aq)$ →	$CaF_2(s)$
Grams at start	29.0	33.0	0
Molar mass, g/mol	164.10	41.99	78.08
Moles at start	0.177	0.786	0

Now you need to identify the limiting reactant. Determine the number of moles of one of the reactants needed to completely react with the other, and state which reactant is limiting. We will show both calculations.

$$0.177 \text{ mol } Ca(NO_3)_2 \times \frac{2 \text{ mol NaF}}{1 \text{ mol } Ca(NO_3)_2} = 0.354 \text{ mol NaF}$$

$$0.786 \text{ mol NaF} \times \frac{1 \text{ mol } Ca(NO_3)_2}{2 \text{ mol NaF}} = 0.393 \text{ mol } Ca(NO_3)_2$$

$Ca(NO_3)_2$ is the limiting reactant.

Fill in the three additional lines in your table: moles used (+), produced (–); moles at end; and grams at end. Do the calculations needed to complete the table.

	$Ca(NO_3)_2(aq)$ +	$2\ NaF(aq)$ →	$CaF_2(s)$
Grams at start	29.0	33.0	0
Molar mass, g/mol	164.10	41.99	78.08
Moles at start	0.177	0.786	0
Moles used (+), produced (–)	– 0.177	– 0.354	+ 0.177
Moles at end	0	0.432	0.177
Grams at end	0	18.1	13.8

Finish by summarizing the answers.

13.8 g CaF_2 will precipitate. NaF is the reactant in excess; 18.1 g NaF is left over.

EXAMPLE 9.17-WB

A solution that contains 29.0 g of calcium nitrate is added to a solution that contains 33.0 g of sodium fluoride. Calculate the number of grams of calcium fluoride that will precipitate. Which reactant is in excess? How many grams of excess reactant are left over?

Write and balance the equation for this double-replacement precipitation reaction.

$Ca(NO_3)_2(aq) + 2\ NaF(aq) \rightarrow CaF_2(s) + 2\ NaNO_3(aq)$

You will need to perform three stoichiometry three-step calculations to complete this problem. The first one is to find the number of grams of calcium fluoride that will precipitate from 29.0 g calcium nitrate, assuming excess sodium fluoride. *PLAN* this calculation and find the answer to this sub-question.

GIVEN: 29.0 g $Ca(NO_3)_2$ *WANTED:* g CaF_2

PER/PATH: g $Ca(NO_3)_2$ $\xrightarrow{164.10\text{ g }Ca(NO_3)_2/\text{mol }Ca(NO_3)_2}$ mol $Ca(NO_3)_2$ $\xrightarrow{1\text{ mol }CaF_2/1\text{ mol }Ca(NO_3)_2}$ mol CaF_2 $\xrightarrow{78.08\text{ mol }CaF_2/\text{mol }CaF_2}$ g CaF_2

$$29.0\text{ g }Ca(NO_3)_2 \times \frac{1\text{ mol }Ca(NO_3)_2}{164.10\text{ g }Ca(NO_3)_2} \times \frac{1\text{ mol }CaF_2}{1\text{ mol }Ca(NO_3)_2} \times \frac{78.08\text{ g }CaF_2}{\text{mol }CaF_2} = 13.8\text{ g }CaF_2$$

Now reverse the assumption. How many grams of calcium fluoride will precipitate from 33.0 g sodium fluoride if there is excess calcium nitrate?

GIVEN: 33.0 g NaF *WANTED:* g CaF_2

PER/PATH: g NaF $\xrightarrow{41.99\text{ g NaF/mol NaF}}$ mol NaF $\xrightarrow{1\text{ mol }CaF_2/2\text{ mol NaF}}$ mol CaF_2 $\xrightarrow{78.08\text{ g }CaF_2/\text{mol }CaF_2}$ g CaF_2

$$33.0\text{ g NaF} \times \frac{1\text{ mol NaF}}{41.99\text{ g NaF}} \times \frac{1\text{ mol }CaF_2}{2\text{ mol NaF}} \times \frac{78.08\text{ g }CaF_2}{\text{mol }CaF_2} = 30.7\text{ g }CaF_2$$

Fill in the blanks in the following sentences:

The limiting reactant is ______________________.

The excess reactant is ______________________.

The number of grams of calcium fluoride that will precipitate is ______________.

The limiting reactant is $Ca(NO_3)_2$. The excess reactant is NaF.

The number of grams of calcium fluoride that will precipitate is 13.8 g.

The calculation from the given 29.0 g $Ca(NO_3)_2$ yielded 13.8 g CaF_2, which was smaller than the mass of CaF_2 that resulted from the assumption that NaF was limiting.

The final stoichiometry problem that needs to be solved is to determine the number of grams of NaF required to react with the mass of limiting reactant, 29.0 g $Ca(NO_3)_2$.

GIVEN: 29.0 g $Ca(NO_3)_2$ *WANTED:* g NaF

PER/PATH: $\text{g Ca(NO}_3)_2 \xrightarrow{164.10 \text{ g Ca(NO}_3)_2/\text{mol Ca(NO}_3)_2} \text{mol Ca(NO}_3)_2 \xrightarrow{2 \text{ mol NaF}/1 \text{ mol Ca(NO}_3)_2} \text{mol NaF} \xrightarrow{41.99 \text{ g NaF/mol NaF}} \text{g NaF}$

$$29.0 \text{ g Ca(NO}_3)_2 \times \frac{1 \text{ mol Ca(NO}_3)_2}{164.10 \text{ g Ca(NO}_3)_2} \times \frac{2 \text{ mol NaF}}{1 \text{ mol Ca(NO}_3)_2} \times \frac{41.99 \text{ g NaF}}{\text{mol NaF}} = 14.8 \text{ g NaF}$$

You just determined that 14.8 g NaF is required to react with all of the initial $Ca(NO_3)_2$. There was 33.0 g NaF initially. How much NaF remains?

33.0 g NaF initial − 14.8 g NaF used = 18.2 g NaF remains

EXAMPLE 9.18-WB

Calculate the number of calories, joules, and kilojoules in a 2,000-Calorie diet. (Assume three significant figures.)

Remember that a food Calorie is a thermochemical kilocalorie. The conversion to calories can be done by moving the decimal point and changing to exponential notation.

2,000 Cal = 2,000 kcal = 2,000,000 cal = 2.00×10^6 cal

Now convert calories to joules.

GIVEN: 2.00×10^6 cal *WANTED:* J

PER/PATH: $\text{cal} \xrightarrow{4.184 \text{ J/cal}} \text{J}$

$$2.00 \times 10^6 \text{ cal} \times \frac{4.184 \text{ J}}{\text{cal}} = 8.37 \times 10^6 \text{ J}$$

Finish by changing joule to kilojoules. You can do this by moving the decimal point, which, in this case, is a matter of adjusting the exponent in the exponential.

$8.37 \times 10^6 \text{ J} = 8.37 \times 10^3 \text{ kJ}$

EXAMPLE 9.19-WB

The iron used to make steel originally occurs in the earth as iron oxides. It is mined and then processed into pure iron. One reaction that happens during processing is the reaction of solid iron(III) oxide with gaseous carbon monoxide, which yields solid iron and gaseous carbon dioxide. For each mole of iron(III) oxide that reacts, 26.8 kJ of energy is released. (a) Is the reaction exothermic or endothermic? (b) Write the thermochemical equation for the reaction in two forms.

Answer Part (a).

The reaction is exothermic.
The problem statement says that energy is *released,* and when energy is released, the process is exothermic.

Now complete Part (b).

$Fe_2O_3(s) + 3\ CO(g) \rightarrow 2\ Fe(s) + 3\ CO_2(g) \quad \Delta H = -26.8\ kJ$

$Fe_2O_3(s) + 3\ CO(g) \rightarrow 2\ Fe(s) + 3\ CO_2(g) + 26.8\ kJ$

EXAMPLE 9.20-WB

$\Delta H = -827$ kJ for the reaction of 2 moles of lead(II) sulfide in the reaction that is the first step in recovering lead from lead(II) sulfide, an ore known as galena. Solid lead(II) sulfide reacts with gaseous oxygen to form solid lead(II) oxide and sulfur dioxide gas. How many kilojoules of energy are released in processing 355 grams of lead(II) sulfide in a laboratory-scale performance of the reaction?

Start by writing a balanced thermochemical equation for the reaction described. You can write either form; we'll show both.

$2\ PbS(s) + 3\ O_2(g) \rightarrow 2\ PbO(s) + 2\ SO_2(g) \quad \Delta H = -827\ kJ$

$2\ PbS(s) + 3\ O_2(g) \rightarrow 2\ PbO(s) + 2\ SO_2(g) + 827\ kJ$

PLAN the solution to the problem.

GIVEN: 355 g PbS *WANTED:* kJ

PER/PATH: g PbS $\xrightarrow{\text{1 mol PbS/239.3 g PbS}}$ mol PbS $\xrightarrow{\text{827 kJ/2 mol PbS}}$ kJ

Complete the solution.

$$355 \text{ g PbS} \times \frac{1 \text{ mol PbS}}{239.3 \text{ g PbS}} \times \frac{827 \text{ kJ}}{2 \text{ mol PbS}} = 613 \text{ kJ}$$

EXAMPLE 9.21-WB

What mass of liquid hexane, $C_6H_{14}(\ell)$, must be burned to provide 8.00×10^3 kJ of heat? $\Delta H = -4.41 \times 10^3$ kJ/mol C_6H_{14} burned.

Write the thermochemical equation.

$$2\ C_6H_{14}(\ell) + 19\ O_2(g) \rightarrow 12\ CO_2(g) + 14\ H_2O(\ell) + 8.82 \times 10^3 \text{ kJ}$$

When the equation is balanced with the lowest whole-number coefficients, 2 moles of $C_6H_{14}(\ell)$ are required. Since the given ΔH is -4.41×10^3 kJ *per mole* of C_6H_{14} burned, it must be multiplied by 2 to be correctly added to this equation. Alternatively, you can "force" the coefficient of $C_6H_{14}(\ell)$ to be 1 and use the given ΔH without change:

$$C_6H_{14}(\ell) + \tfrac{19}{2}\ O_2(g) \rightarrow 6\ CO_2(g) + 7\ H_2O(\ell) + 4.41 \times 10^3 \text{ kJ}$$

Take the remainder of the solution to completion.

GIVEN: 8.00×10^3 kJ *WANTED:* mass C_6H_{14} (assume g)

PER/PATH: kJ $\xrightarrow{8.82 \times 10^3 \text{ kJ/2 mol } C_6H_{14}}$ mol C_6H_{14} $\xrightarrow{86.17 \text{ g } C_6H_{14}\text{/mol } C_6H_{14}}$ g C_6H_{14}

$$8.00 \times 10^3 \text{ kJ} \times \frac{2 \text{ mol } C_6H_{14}}{8.82 \times 10^3 \text{ kJ}} \times \frac{86.17 \text{ g } C_6H_{14}}{\text{mol } C_6H_{14}} = 156 \text{ g } C_6H_{14}$$

Name: ______________________ Date: ______________________

ID: ______________________ Section: ______________________

Questions, Exercises, and Problems

Section 9.1: Conversion Factors from a Chemical Equation

1. The first step in the Ostwald process for manufacturing nitric acid is the reaction between ammonia and oxygen described by the equation $4\ NH_3 + 5\ O_2 \rightarrow 4\ NO + 6\ H_2O$. This equation is used to answer all parts of this question.

   a) How many moles of ammonia will react with 95.3 moles of oxygen?

   Answer: ☐

   b) How many moles of nitrogen monoxide will result from the reaction of 2.89 moles of ammonia?

   Answer: ☐

   c) If 3.35 moles of water is produced, how many moles of nitrogen monoxide will also be produced?

   Answer: ☐

2. Magnesium hydroxide is formed from the reaction of magnesium oxide and water. How many moles of magnesium oxide are needed to form 0.884 mole of magnesium hydroxide, when the oxide added to excess water?

   Answer: ☐

Name: ______________________ Date: ______________________

ID: ______________________ Section: ______________________

3. When sulfur dioxide reacts with oxygen, sulfur trioxide is formed. How many moles of sulfur dioxide are needed to produce 3.99 moles of sulfur trioxide if the reaction is carried out in excess oxygen?

Answer: [ ]

Section 9.2: Mass Calculations

4. The first step in the Ostwald process for manufacturing nitric acid is the reaction between ammonia and oxygen described by the equation $4\ NH_3 + 5\ O_2 \rightarrow 4\ NO + 6\ H_2O$. This equation is used to answer all parts of this question.

a) How many moles of ammonia can be oxidized by 268 grams of oxygen?

Answer: [ ]

b) If the reaction consumes 31.7 moles of ammonia, how many grams of water will be produced?

Answer: [ ]

c) How many grams of ammonia are required to produce 404 grams of nitrogen monoxide?

Answer: [ ]

Name: ______________________ Date: ______________________

ID: ______________________ Section: ______________________

d) If 6.41 grams of water result from the reaction, what will be the yield of nitrogen monoxide (in grams)?

Answer: [ ]

5. The explosion of nitroglycerine is described by the equation $4\ C_3H_5(NO_3)_3 \rightarrow 12\ CO_2 + 10\ H_2O + 6\ N_2 + O_2$. How many grams of carbon dioxide are produced by the explosion of 21.0 grams of nitroglycerine?

Answer: [ ]

6. Soaps are produced by the reaction of sodium hydroxide with naturally occurring fats. The equation for one such reaction is $C_3H_5(C_{17}H_{35}COO)_3 + 3\ NaOH \rightarrow C_3H_5(OH)_3 + 3\ C_{17}H_{35}COONa$. The last compound is the soap. Calculate the number of grams of sodium hydroxide required to produce 323 grams of soap by this method.

Answer: [ ]

7. One way to make sodium thiosulfate, known as "hypo" and used in photographic developing, is described by the equation $Na_2CO_3 + 2\ Na_2S + 4\ SO_2 \rightarrow 3\ Na_2S_2O_3 + CO_2$. How many grams of sodium carbonate are required to produce 681 grams of sodium thiosulfate?

Answer: [ ]

Name: ______________________ Date: ______________________

ID: ______________________ Section: ______________________

8. The hard water scum that forms a ring around the bathtub is an insoluble soap, $Ca(C_{18}H_{35}O_2)_2$. It is formed when a soluble soap, $NaC_{18}H_{35}O_2$, reacts with the calcium ion that is responsible for the hardness in water: $2\ NaC_{18}H_{35}O_2 + Ca^{2+} \rightarrow Ca(C_{18}H_{35}O_2)_2 + 2\ Na^+$. How many milligrams of scum can be formed from 616 milligrams of $NaC_{18}H_{35}O_2$?

Answer: [ ]

9. Pig iron from a blast furnace contains several impurities, one of which is phosphorus. Additional iron ore, Fe_2O_3, is included with pig iron in making steel. The oxygen in the ore oxidizes the phosphorus by the reaction $12\ P + 10\ Fe_2O_3 \rightarrow 3\ P_4O_{10} + 20\ Fe$. If a sample of the remains from the furnace contains 802 milligrams of tetraphosphorus decoxide, how many grams of Fe_2O_3 was used in making it?

Answer: [ ]

Sodium carbonate, Na_2CO_3, also known as washing soda, and sodium hydrogen carbonate, $NaHCO_3$, better known as baking soda, are two very important industrial chemicals that can be made by the Solvay process. The next two questions are based on reactions in that process.

10. How much NaCl is needed to react completely with 83.0 grams of ammonium hydrogen carbonate in $NaCl + NH_4HCO_3 \rightarrow NaHCO_3 + NH_4Cl$?

Answer: [ ]

Name: ______________________ Date: ______________________

ID: ______________________ Section: ______________________

11. By-product ammonia is recovered from the Solvay process by the reaction $Ca(OH)_2$ + $2\ NH_4Cl \rightarrow CaCl_2 + 2\ H_2O + 2\ NH_3$. How many grams of calcium chloride can be produced along with 60.2 grams of ammonia?

Answer: [ ]

12. How many grams of sodium hydroxide are needed to neutralize completely 32.6 grams of phosphoric acid?

Answer: [ ]

13. What mass of magnesium hydroxide will precipitate if 2.09 grams of potassium hydroxide is added to a magnesium nitrate solution?

Answer: [ ]

Name: ______________________ Date: ______________________

ID: ______________________ Section: ______________________

14. An experimenter recovers 0.521 gram of sodium sulfate from the neutralization of sodium hydroxide by sulfuric acid. How many grams of sulfuric acid reacted?

Answer:

15. The reaction of a dry cell may be represented as follows: zinc reacts with ammonium chloride to form zinc chloride, ammonia, and hydrogen. Calculate the number of grams of zinc consumed during the release of 7.05 grams of ammonia in such a cell.

Answer:

Section 9.3: Gas Stoichiometry at Standard Temperature and Pressure (Optional)

16. Explain the restrictions placed on the statement that 22.4 L/mol is the molar volume of any gas.

Answer:

Name: ____________________ Date: ____________________

ID: ____________________ Section: ____________________

17. What volume will be occupied by 4.21 moles of carbon monoxide at STP?

Answer: [ ]

18. Sulfur dioxide, used in making sulfuric acid, can be obtained from sulfide ores by the reaction $4\ FeS_2(s) + 11\ O_2(g) \rightarrow 2\ Fe_2O_3(s) + 8\ SO_2(g)$. Calculate the grams of FeS_2 that must react to produce 423 L SO_2, measured at STP.

Answer: [ ]

19. How many liters of hydrogen, measured at STP, will result from the electrolytic decomposition of 12.6 grams of water?

Answer: [ ]

20. In the natural oxidation of hydrogen sulfide released by decaying organic matter, the following reaction occurs: $2\ H_2S(g) + 3\ O_2(g) \rightarrow 2\ SO_2(g) + 2\ H_2O(g)$. How many milliliters of oxygen are required to react with 2.09 L of hydrogen sulfide if both gases are measured at STP?

Answer: [ ]

Name: ______________________ Date: ______________________

ID: ______________________ Section: ______________________

Section 9.4: Gas Stoichiometry at Nonstandard Conditions (Optional)

21. One source of sulfur dioxide used in making sulfuric acid comes from sulfide ores by the reaction $4\ FeS_2(s) + 11\ O_2(g) \rightarrow 2\ Fe_2O_3(s) + 8\ SO_2(g)$. How many liters of $SO_2(g)$, measured at 983 torr and 214°C, are produced by the reaction of 598 g FeS_2?

Answer: ______________________

22. How many grams of water must decompose by electrolysis to produce 23.9 L H_2, measured at 28°C and 728 torr?

Answer: ______________________

23. The reaction chamber in a modified Haber process for making ammonia by the direct combination of its elements is operated at 575°C and 248 atm. How many liters of nitrogen, measured at these conditions, will react to produce 9.16×10^3 g NH_3?

Answer: ______________________

Name: ______________________ Date: ______________________

ID: ______________________ Section: ______________________

24. When properly detonated, ammonium nitrate explodes violently, releasing hot gases: $NH_4NO_3(s) \rightarrow N_2O(g) + 2\ H_2O(g)$. If the total volume of gas produced, both dinitrogen oxide and steam, is 82.3 L at 447°C and 896 torr, how many grams of NH_4NO_3 exploded?

Answer: [ ]

Section 9.5: Percent Yield

25. The function of "hypo" (see Question 7) in photographic developing is to remove excess silver bromide by the reaction $2\ Na_2S_2O_3 + AgBr \rightarrow Na_2Ag(S_2O_3)_2 + NaBr$. What is the percent yield if the reaction of 8.18 grams of sodium thiosulfate produces 2.61 grams of sodium bromide?

Answer: [ ]

26. Calcium cyanamide is a common fertilizer. When mixed with water in the soil, it reacts to produce calcium carbonate and ammonia: $CaCN_2 + 3\ H_2O \rightarrow CaCO_3 + 2\ NH_3$. How much ammonia can be obtained from 7.25 grams of calcium cyanamide in a laboratory experiment in which the percent yield will be 92.8%?

Answer: [ ]

Name: ______________________ Date: ______________________

ID: ______________________ Section: ______________________

27. The Haber process for making ammonia from the nitrogen of the air is given by the equation $N_2 + 3\ H_2 \rightarrow 2\ NH_3$. Calculate the mass of hydrogen that must be supplied to make 5.00×10^2 kg of ammonia in a system that has an 88.8% yield.

Answer: [ ]

28. Calculate the percent yield in the photosynthesis reaction by which carbon dioxide is converted to sugar if 7.03 grams of carbon dioxide yields 3.92 grams of $C_6H_{12}O_6$. The equation is $6\ CO_2 + 6\ H_2O \rightarrow C_6H_{12}O_6 + 6\ O_2$.

Answer: [ ]

29. Ethylacetate, $CH_3COOC_2H_5$, is manufactured by the reaction between acetic acid, CH_3COOH, and ethanol, C_2H_5OH: $CH_3COOH + C_2H_5OH \rightarrow CH_3COOC_2H_5 + H_2O$. How much acetic acid must be used to get 62.5 kilograms of ethylacetate if the percent yield is 69.1%?

Answer: [ ]

Name: ______________________ Date: ______________________

ID: ______________________ Section: ______________________

Sections 9.6–8: Limiting Reactants

30. A solution containing 1.63 grams of barium chloride is added to a solution containing 2.40 grams of sodium chromate (chromate ion, CrO_4^{2-}). Find the number of grams of barium chromate that can precipitate. Also determine which reactant was in excess, as well as the number of grams over the amount required by the limiting reactant.

Answer: ______________________

31. The equation for one method of preparing iodine is $2\ NaIO_3 + 5\ NaHSO_3 \rightarrow I_2 + 3\ NaHSO_4 + 2\ Na_2SO_4 + H_2O$. If 6.00 kilograms of sodium iodate is reacted with 7.33 kilograms of sodium hydrogen sulfite, how many kilograms of iodine can be produced? Which reactant will be left over? How many kilograms will be left?

Answer: ______________________

Name: ______________________ Date: ______________________

ID: ______________________ Section: ______________________

32. A mixture of tetraphosphorus trisulfide and powdered glass is in the white tip of strike-anywhere matches. The P_4S_3 is made by the direct combination of the elements: $8\ P_4 + 3\ S_8 \rightarrow 8\ P_4S_3$. If 133 grams of phosphorus is mixed with the full contents of a 126-g (4-oz) bottle of sulfur, how many grams of the compound can be formed? How much of which element will be left over?

Answer: [ ]

Section 9.9: Energy

33. a) How many kilojoules are equal to 0.731 kcal?

Answer: [ ]

b) What number of calories is the same as 651 J?

Answer: [ ]

c) Determine the number of kilocalories that is equivalent to 6.22×10^3 J.

Answer: [ ]

Name: ______________________ Date: ______________________

ID: ______________________ Section: ______________________

34. When 15 g of carbon is burned, 493 kJ of energy is released. Calculate the number of calories and kilocalories this represents.

Answer: [ ]

35. Each day, 5.8×10^2 kcal of heat is removed from the body by evaporation. How many kilojoules are removed in a year?

Answer: [ ]

Section 9.10: Thermochemical Equations

Questions 36 through 38: Thermochemical equations may be written in two ways, either with an energy term as a part of the equation or with ΔH set apart from the regular equation. In the questions that follow, write both forms of the equations for the reactions described. Recall that state designations are required for all substances in a thermochemical equation.

36. Energy is absorbed from sunlight in the photosynthesis reaction in which carbon dioxide and water vapor combine to produce sugar, $C_6H_{12}O_6$, and release oxygen. The amount of energy is 2.82×10^3 kJ per mole of sugar formed.

Answer: [ ]

37. The electrolysis of water is an endothermic reaction, absorbing 286 kJ for each mole of liquid water decomposed to its elements.

Answer: [ ]

Name: ______________________ Date: ______________________

ID: ______________________ Section: ______________________

38. The reaction in an oxyacetylene torch is highly exothermic, releasing 1.31×10^3 kJ of heat for every mole of acetylene, $C_2H_2(g)$, burned. The end products are gaseous carbon dioxide and liquid water.

Answer: [ ]

Section 9.11: Thermochemical Stoichiometry

39. Quicklime, the common name for calcium oxide, CaO, is made by heating limestone, $CaCO_3$, in slowly rotating kilns about $2\frac{1}{2}$ meters in diameter and about 60 meters long. The reaction is $CaCO_3(s) + 178\text{ kJ} \rightarrow CaO(s) + CO_2(g)$. How many kilojoules are required to decompose 5.80 kg of limestone?

Answer: [ ]

40. The quicklime produced in Question 39 is frequently converted to calcium hydroxide, sometimes called slaked lime, by an exothermic reaction with water: $CaO(s) + H_2O(\ell) \rightarrow Ca(OH)_2(s) + 65.3\text{ kJ}$. How many grams of quicklime was processes in a reaction that produced 291 kJ of energy?

Answer: [ ]

Name: ______________________ Date: ______________________

ID: ______________________ Section: ______________________

41. How many grams of octane, a component of gasoline, would you have to use in your car to produce 9.48×10^5 kJ of energy? $\Delta H = 1.09 \times 10^4$ kJ for the reaction $2\ C_8H_{18}(\ell) + 25\ O_2(g) \rightarrow 16\ CO_2(g) + 18\ H_2O(\ell)$.

Answer: [ ]

General Questions

42. How many grams of calcium phosphate will precipitate if excess calcium nitrate is added to a solution containing 3.98 grams of sodium phosphate?

Answer: [ ]

43. Baking cakes and pastries involves the production of CO_2 to make the batter rise. For example, citric acid, $H_3C_6H_5O_7$, in lemon or orange juice can react with baking soda, $NaHCO_3$ to produce carbon dioxide gas: $H_3C_6H_5O_7(aq) + 3\ NaHCO_3(aq) \rightarrow Na_3C_6H_5O_7(aq) + 3\ CO_2(g) + 3\ H_2O(\ell)$.

a) If 6.00 g $H_3C_6H_5O_7$ react with 20.0 g $NaHCO_3$, how many grams of carbon dioxide will be produced?

Answer: [ ]

Name: ____________________ Date: ____________________

ID: ____________________ Section: ____________________

b) How many grams of which reactant will remain unreacted?

Answer: [ ]

c) Can you name $Na_3C_6H_5O_7$? Remember, it comes from citr*ic* acid.

44. How much energy is required to decompose 1.42 grams of $KClO_3$ according to the following equation: $2\ KClO_3(s) \rightarrow 2\ KCl(s) + 3\ O_2(g)$? $\Delta H = 89.5$ kJ for the reaction.

Answer: [ ]

More Challenging Problems

45. Carborundum, SiC, is widely used as an abrasive in industrial grinding wheels. It is prepared by the reaction of sand, SiO_2, with the carbon in coke: $SiO_2 + 3\ C \rightarrow SiC + 2\ CO$. How many kilograms of carborundum can be prepared from 727 kg of coke that is 88.9% carbon?

Answer: [ ]

Name: ______________________ Date: ______________________

ID: ______________________ Section: ______________________

46. The chemical equation that describes what happens in an automobile storage battery as it generates electrical energy is $PbO_2 + Pb + 2\ H_2SO_4 \rightarrow 2\ PbSO_4 + 2\ H_2O$.

a) What fraction of the lead in $PbSO_4$ comes from PbO_2 and what fraction comes from elemental lead?

Answer: [ ]

b) If the operation of the process uses 29.7 g $PbSO_4$, how many grams of PbO_2 must have been consumed?

Answer: [ ]

47. A dry mixture of hydrogen chloride and air is passed over a heated catalyst in the Deacon process for manufacturing chlorine. Oxidation occurs by the following reaction: $4\ HCl + O_2 \rightarrow 2\ Cl_2 + 2\ H_2O$. If the conversion is 63% complete, how many tons of chlorine can be recovered from 1.4 tons of HCl? (*Hint:* Whatever you can do with moles, kilomoles, and millimoles, you can also do with ton-moles.)

Answer: [ ]

Name: ______________________ Date: ______________________

ID: ______________________ Section: ______________________

48. In the recovery of silver from silver chloride waste (see textbook Question 48) a certain quantity of waste material is estimated to contain 184 g of silver chloride, AgCl. The treatment tanks are charged with 45 grams of zinc and 145 grams of sodium cyanide, NaCN. Is there enough of the two reactants to recover all of the silver from the AgCl? If no, how many grams of silver chloride will remain? If yes, how many more grams of silver chloride could have been treated by the available Zn and NaCN?

Answer: [ ]

49. Nitroglycerine is the explosive ingredient in industrial dynamite. Much of its destructive force comes from the sudden creation of large volumes of gaseous products. A great deal of energy is released, too. $\Delta H = -6.17 \times 10^3$ kJ for the equation $4\ C_3H_5(NO_3)_3(\ell) \rightarrow 12\ CO_2(g) + 10\ H_2O(g) + 6\ N_2(g) + O_2(g)$. Calculate the number of pounds of nitroglycerine that must be used in a blasting operation that requires 5.88×10^4 kJ of energy.

Answer: [ ]

Name: ______________________ Date: ______________________

ID: ______________________ Section: ______________________

50. A researcher dissolved 1.382 grams of impure copper were dissolved in nitric acid to produce a solution of $Cu(NO_3)_2$. The solution went through a series of steps in which $Cu(NO_3)_2$ was changed to $Cu(OH)_2$, then to CuO, and then to a solution of $CuCl_2$. This was treated with an excess of a soluble phosphate, precipitating all the copper in the original sample as pure $Cu_3(PO_4)_2$. The precipitate was dried and weighed. Its mass was 2.637 g. Find the percent copper in the original sample.

Answer: [ ]

51. If a solution of silver nitrate, $AgNO_3$, is added to a second solution containing a chloride, bromide, or iodide, the silver ion, Ag^+, from the first solution will precipitate the halide as silver chloride, silver bromide, or silver iodide. If excess $AgNO_3(aq)$ is added to a mixture of the above halides, it will precipitate them both, or all, as the case may be. A solution contains 0.230 g NaCl and 0.771 g NaBr. What is the smallest quantity of $AgNO_3$ that is required to precipitate both halides completely?

Answer: [ ]

Answers to Target Checks

1. (a) Three moles of CD result. One mole of D remains.
(b) Two moles of EF_2 result. One mole of E remains.
(c) The reaction stoichiometry is different; the limiting reactant depends on both the number of moles of reactants and the stoichiometric ratio by which they react.

Chapter 10

Atomic Theory: The Quantum Model of the Atom

☑ TARGET CHECK 10.1-WB

a) Give a brief word description of an atom according to the Bohr model.

b) Explain why the spectral lines are discontinuous when the light from an elemental gas discharge tube is passed through a prism.

c) Are the electrons in the gas contained in an elemental gas discharge tube in the ground state or in an excited state when no electricity is passed through the tube? Explain.

☑ TARGET CHECK 10.2-WB

a) What are the principal energy levels in an atom? Arrange them from lowest energy to highest energy.

b) List the sublevels for level $n = 4$ in order of increasing energy.

c) Sketch all the *s* and *p* orbitals for level $n = 3$.

d) How many *d* orbitals are there for level $n = 4$?

e) An electron orbital may be occupied by 1 or 2 electrons, but by no other number. True or false? If true, why is it true? If false, explain why it is false.

EXAMPLE 10.1-WB

a) List the following electron sublevels in order of increasing energy: $3s$ $3p$ $3d$ $4s$ $4p$

_________ < _________ < _________ < _________ < _________

b) Write the electron configuration of the highest occupied energy sublevel for each of the following elements:

sodium _________ carbon _________ nickel _________

---

a) $3s < 3p < 4s < 3d < 4p$
Reading the periodic table top to bottom and left to right, as illustrated in Figure 10.8, we see that Period 3 has the $3s$ and $3p$ blocks, and then Period 4 has the $4s$, $3d$, and $4p$ blocks, from left to right. This is the order of increasing energy.

b) sodium: $3s^1$ carbon: $2p^2$ nickel: $3d^8$
When you first locate sodium, Na, on the periodic table, you see that it is in the *s*-block (Groups 1A/1 and 2A/2). You then count down to the Na box and see that it is in the third period, thus its highest occupied energy sublevel is $3s$. It is the first leftmost element in the *s*-block, so its configuration is $3s^1$. Following a similar logic for carbon, C, it is in the *p*-block (Groups 3A/13 through 8A/18) and the second period, so the sublevel is $2p$. Start counting at B, boron, and C, carbon, is the second element in the block, so its configuration is $2p^2$. For nickel, it is in the *d*-block (U.S. B groups or IUPAC groups 3–12) and the fourth period, so it is a $3d$ element (remember that the *d*-block starts with $n = 3$). When you count left to right across the $3d$ row, beginning with Sc, you see that Ni is the eighth element, so its configuration is $3d^8$.

EXAMPLE 10.2-WB

Write the complete electron configuration for an atom of sulfur, and then rewrite its configuration with a Group 8A/18 core.

Recall the elemental symbol for sulfur and locate it on the periodic table. Then identify the electron configuration of the highest occupied energy sublevel.

$3p$ (sulfur is S, Z = 16)

Now list all sublevels of lower energy than the $3p$ sublevel, in order of increasing energy. To do so, read the periodic table, line by line, from left to right.

$1s$ $2s$ $2p$ $3s$ $3p$

Add the superscripts to complete your electron configuration. Be sure that their sum is equal to the number of electrons in a neutral sulfur atom.

$1s^22s^22p^63s^23p^4$ $2 + 2 + 6 + 2 + 4 = 16$ ✓

Finally, rewrite the electron configuration with a Group 8A/18 core.

[Ne]$3s^2 3p^6$

EXAMPLE 10.3-WB

Write the complete electron configuration and an electron configuration with a Group 8A/18 core for the two elements among those with atomic numbers 1 through 36 that do not follow the "expected" filling order.

What are the two elements among those with atomic numbers 1 through 36 that do not follow the standard filling order? Write their names and symbols. Locate their positions on the periodic table, and write their atomic numbers.

Chromium, Cr, Z = 24 and copper, Cu, Z = 29

What is the "expected" ground-state electron configuration for chromium? Use a Group 8A/18 core.

[Ar]$4s^2 3d^4$

The Group 8A/18 element with the highest atomic number less than that of chromium is argon, Z = 18, so we have an argon core, [Ar]. The *s*-block elements that follow are in the fourth period, so the next sublevel to be filled is 4*s*, and there are two elements in this block, so we have [Ar]$4s^2$. In the 3*d* block that follows, counting from left to right requires four electrons to get to Cr, yielding [Ar]$4s^2 3d^4$.

There is a special energetic stability associated with a complete set of half-filled orbitals, and there are five *d* orbitals available at every principal energy level that has *d* orbitals. Our incorrect configuration for chromium shows one electron in four of the five available *d* orbitals, which leaves one unoccupied orbital. The electron from the next lowest energy sublevel is promoted to fill the fifth *d* orbital. Write the correct electron configuration for chromium.

[Ar]$4s^1 3d^5$

An electron that would otherwise occupy the 4*s* sublevel is placed in the 3*d* sublevel to achieve a half-filled configuration, one with one electron in each of the five *d* orbitals.

Now write the electron configuration that copper would have according to the "expected" filling order. Use a noble gas core.

[Ar]$4s^2 3d^9$

Copper is the ninth element in the 3*d* block.

As with half-filled orbitals, there is an energetic stability associated with a complete set of filled orbitals. Nine *d* electrons in five orbitals means that four orbitals are filled with two electrons, and one has just one electron. An electron is therefore promoted from the next lowest energy sublevel to give ten *d* electrons, filling each orbital. Write the actual electron configuration for copper.

$[Ar]4s^1 3d^{10}$

The $3d^{10}$ configuration fills all five of the *d* orbitals in that sublevel.

Complete the example by writing the complete (no Group 8A/18 core) electron configuration for chromium and copper. For each, verify that the total number of electrons in your configuration is equal to the number in the atom.

Cr: $1s^2 2s^2 2p^6 3s^2 3p^6 4s^1 3d^5$ $2 + 2 + 6 + 2 + 6 + 1 + 5 = 24$ ✓

Cu: $1s^2 2s^2 2p^6 3s^2 3p^6 4s^1 3d^{10}$ $2 + 2 + 6 + 2 + 6 + 1 + 10 = 29$ ✓

EXAMPLE 10.4-WB

Refer to a periodic table, and state the number of valence electrons you expect for an atom of (a) francium (Z = 87) and (b) radium (Z = 88). Using *n* for the highest occupied energy level, write the configuration of the valence electrons for each element.

In which group of the periodic table do you find francium? What is the group number for radium?

Fr, Group 1A/1 Ra, Group 2A/2

What is the number of valence electrons for *all* elements in Group 1A/1? How many valence electrons do Group 2A/2 elements have? What is the number of valence electrons for each of these elements?

Group 1A/1 elements have 1 valence electron. Group 2A/2 elements have 2 valence electrons. Francium therefore has 1 valence electron, and radium has 2 valence electrons.

What is the general electron configuration for all Group 1A/1 elements? What is the general configuration for the Group 2A/2 elements? Use the general form $ns^x np^y$.

Group 1A/1: ns^1 Group 2A/2: ns^2

Can you write the actual electron configuration for francium and radium? Try it, using a Group 8A/18 core.

Fr: [Rn]$7s^1$ Ra: [Rn]$7s^2$

The highest atomic-numbered Group 8A/18 element that is less than the atomic number of francium and radium is radon, Rn, Z = 86. Both elements are in the *s*-block and the seventh period.

EXAMPLE 10.5-WB

(a) What are the minimum and maximum numbers of valence electrons an atom can have? Explain.
(b) Write the Lewis symbols for the Period 2 elements with the minimum and maximum number of valence electrons.

Answer Part (a) first.

An atom may have as few as 1 valence electron (Group 1A/1, ns^1) to as many as 8 (Group 8A/18, ns^2np^6).

Now write the Lewis symbols of the second period elements with one and eight valence electrons.

Li ·   :Ne: (with two dots above and two dots below Ne)

EXAMPLE 10.6-WB

(a) Figure 10.14 shows that a magnesium atom is smaller than a calcium atom. Explain why this is true.
(b) Figure 10.14 shows that a magnesium atom is larger than an aluminum atom. Explain why this is true.

Part (a) asks you to explain the periodic trend of increasing size down the table in any group. Refer to the text immediately following Goal 16, if necessary, to review this concept.

The highest occupied principal energy level in a magnesium atom is $n = 3$, and in a calcium atom, it is $n = 4$. As valence electrons occupy higher principal energy levels, they are generally farther from the nucleus, and the atoms become larger.

Part (b) asks about the reason for the periodic trend across a period. Write your explanation.

The highest occupied principal energy level in both magnesium and aluminum atoms is $n = 3$, so this variable is the same in each case. However, aluminum (Z = 13) has a greater nuclear charge than magnesium (Z = 12). As the number of protons in an atom increases, the positive charge in the nucleus increases. This pulls the valence electrons closer to the nucleus, so the atom becomes smaller.

☑ TARGET CHECK 10.3-WB

a) List the following elements in order of increasing first ionization energy: Na, S, Ar.
b) Explain why magnesium, calcium, and barium are all members of the same chemical family.
c) List the atomic numbers of the members of the halogen family.
d) List atoms of atomic numbers 6, 7, 14, and 32 in order of increasing size.
e) List the symbols of the nonmetals in Period 4.

Name: ______________________ Date: ______________________

ID: ______________________ Section: ______________________

Questions, Exercises, and Problems

Section 10.1: The Bohr Model of the Hydrogen Atom

1. The visible spectrum is a small part of the whole electromagnetic spectrum. Name several other parts of the whole spectrum that are included in our everyday vocabulary.

2. Identify measurable wave properties that are used in describing light.

3. Which among the following are *not* quantized? (a) shoelaces; (b) cars passing through a toll plaza in a day; (c) birds in an aviary; (d) flow of a river in m^3/hr; (e) percentage of salt in a solution.

4. What kind of light would be emitted by atoms if the electron energy were not quantized? Explain.

5. What must be done to an atom, or what must happen to an atom, before it can emit light?

6. Which atom is more apt to emit light, one in the ground state or one in an excited state? Why?

Name: ______________________ Date: ______________________

ID: ______________________ Section: ______________________

7. Using a sketch of the Bohr model of an atom, explain the source of the observed lines in the spectrum of hydrogen.

8. Identify the major advances that came from the Bohr model of the atom.

Section 10.2: The Quantum Mechanical Model of the Atom

9. Compare the relative energies of the principal energy levels within the same atom.

10. How many sublevels are present in each principal energy level?

11. How many orbitals are in the *s* sublevel? The *p* sublevel? The *d* sublevel? The *f* sublevel?

Name: ______________________ Date: ______________________

ID: ______________________ Section: ______________________

12. Each p sublevel contains six orbitals. Comment on this statement.

13. The principal energy level with $n = 6$ contains six sublevels, although not all six are occupied for any element now known. Comment on this statement.

14. An orbital may hold one electron or two electrons, but no other number. Comment on this statement.

15. What is your opinion of the common picture showing one or more electrons whirling around the nucleus of an atom?

16. What is the largest number of electrons that can occupy the $4p$ orbitals? the $3d$ orbitals? the $5f$ orbitals?

17. Energies of the principal energy levels of an atom sometimes overlap. Explain how this is possible.

Name: ______________________ Date: ______________________

ID: ______________________ Section: ______________________

18. Which of the following statements is true? The quantum mechanical model of the atom includes (a) all of the Bohr model; (b) part of the Bohr model; (c) no part of the Bohr model. Justify your answer.

Section 10.3: Electron Configuration

19. What do you conclude about the symbol $2d^5$? (Careful. . . .)

20. What element has the electron configuration $1s^2 2s^2 2p^6 3s^2 3p^4$? In which period and group of the periodic table is it located?

21. What is the argon core? What is its symbol, and what do you use it for?

Questions 22 and 23: Identify the elements whose electron configurations are given.

22. a) $1s^2 2s^2 2p^6 3s^2 3p^6 4s^2 3d^{10} 4p^4$

b) $1s^2 2s^2 2p^4$

c) $1s^2 2s^2 2p^6 3s^2$

23. a) $[He]2s^2 2p^3$

b) $[Ne]3s^2 3p^1$

c) $[Ar]4s^2 3d^7$

Name: ____________________ Date: ____________________

ID: ____________________ Section: ____________________

Questions 24 and 25: Write complete ground-state electron configurations of an atom of the elements shown. Do not use a noble gas core.

24. Magnesium and nickel

25. Chromium and selenium (Z = 34)

26. If it can be done, rewrite the electron configurations in Questions 24 and 25 with a neon or argon core.

27. Use a noble gas core to write the electron configuration of vanadium (Z = 23).

Section 10.4: Valence Electrons

28. Why are valence electrons important?

29. What are the valence electrons of aluminum? Write both ways they may be represented.

Name: ______________________ Date: ______________________

ID: ______________________ Section: ______________________

30. Using n for the principal quantum number, write the electron configuration for the valence electrons of Group 6A/16 atoms.

Section 10.5: Trends in the Periodic Table

31. What is the general trend in ionization energies of elements in the same chemical family?

32. Suggest an explanation for the general trend of increasing ionization energy across a period.

33. What is it about strontium (Z = 38) and barium that makes them members of the same chemical family?

34. To what family does the electron configuration ns^2np^5 belong?

35. Expressed as ns^xnp^y, what electron configuration is isoelectronic with a noble gas?

Name: ______________________ Date: ______________________

ID: ______________________ Section: ______________________

36. Identify the chemical families in which (a) krypton (Z = 36) and (b) beryllium are found.

37. Account for the chemical similarities between chlorine and iodine in terms of their electron configurations.

38. Where in the periodic table do you find the transition elements? Are they metals or nonmetals?

39. Identify an atom in Period 4 that is (a) larger than an atom or arsenic (Z = 33) and (b) smaller than an atom of arsenic.

40. Why does atomic size decrease as you go left to right across a row in the periodic table?

41. Even though an atom of germanium (Z = 32) has more than twice the nuclear charge of an atom of silicon (Z = 14), the germanium atom is larger. Explain.

Name: ______________________ Date: ______________________

ID: ______________________ Section: ______________________

42. What are *metalloids* or *semimetals*? How do their properties compare with those of metal and nonmetals?

Use the periodic table, Figure 10.16, in the textbook, to answer Questions 43 to 45. Answer with letters (D, E, X, Z, and so on) from the figure.

43. Give the letters that are in the positions of (a) alkaline earth metals and (b) noble gases.

44. List the letters that correspond to nonmetals.

45. List the elements Q, X, and Z in order of increasing atomic size (smallest atom first).

General Questions

46. Figure 10.3 shows the spectra of hydrogen and neon. Hydrogen has four lines, one of which is sometimes difficult to see. Yet we say there are more lines in the hydrogen spectrum. Why do we not see them?

More Challenging Problems

47. Write the electron configurations you would expect for iodine and tungsten ($Z = 74$).

Name: ______________________ Date: ______________________

ID: ______________________ Section: ______________________

48. Why is the definition of atomic number based on the number of protons in an atom rather than the number of electrons?

49. Suggest reasons why the ionization energy of magnesium is greater than the ionization energies of sodium, aluminum, and calcium.

50. Do elements become more metallic or less metallic as you (a) go down a group in the periodic table, (b) move left to right across a period in the periodic table? Support your answers with examples.

51. Although the quantum model of the atom makes predictions for atoms of all elements, most of the quantitative confirmation of these predictions is limited to substances whose formulas are H, He^+, and H_2^+. From your knowledge of electron configurations and the limited information about chemical formulas given in this chapter, can you identify a single feature that all three substances have that makes them unique and probably the easiest substances to investigate?

Name: ____________________ Date: ____________________

ID: ____________________ Section: ____________________

52. Carbon does not form a stable monatomic ion. Suggest a reason for this.

53. Xenon (Z = 54) was the first noble gas to be chemically combined with another element. Xenon and krypton are present in nearly all of the small number of noble gas compounds known today. Note the ionization energy trend begun with the other noble gases in Figure 10.12. What do these facts suggest about the relative reactivities of the noble gases and the character of noble gas compounds?

54. Figure 10.12 indicates a general increase in the first ionization energy from left to right across a row of the periodic table. However, there are two sharp breaks in both shown. (a) Suggest a reason why ionization energy should increase as atomic number increases within a period. (b) Can you correlate features of electron configurations with the locations of the breaks?

Answers to Target Checks

1. a) The Bohr model of the atom describes electrons moving in circular orbits of fixed radii. The electrons have fixed amounts of energy and move between levels in quantum jumps.
b) If energy is added to any gas, atoms of that gas absorb some of the energy by having their electrons lifted from ground state to some higher quantized energy level. These excited state electrons are unstable at their high energy and fall spontaneously in one or more steps to the ground state, emitting energy equal to the energy absorbed. If the emitted energy is in the visible range, it appears as light. This process produces discrete lines through a prism because electron energies are quantized, and only that specific energy between quantized energy levels may be released. Specific energy levels in the visible range result in specific colors.
c) The electrons are in the ground state. When the electricity powering the gas discharge tube is off, no energy is added to the atoms of gas in the tube. If atoms do not absorb energy, their electrons remain in the ground state.

2. a) Principal energy levels are identified by their principal quantum numbers, designated n. The principal quantum numbers are a set of whole numbers beginning with 1. Energy increases as n increases.
b) In order of increasing energy, the sublevels for $n = 4$ are $4s < 4p < 4d < 4f$.
c) See Figure 10.7.
d) There are five d orbitals in each d sublevel for n greater than or equal to three.
e) The statement may be either true or false, depending upon how it is interpreted. An orbital may be thought of as unoccupied, or stated in different words, occupied by zero electrons. Thus the statement can be interpreted as false. Technically, however, an orbital is not literally a "thing" in which an electron is placed. An orbital is a description of where an electron is likely to be found. If there is no electron, there can be no description of where it is likely to be found—no orbital. Thus the statement can be considered to be true.

3. a) Na < S < Ar
b) Magnesium, calcium, and barium share an ns^2 highest energy electron configuration. These two electrons are easily lost to form monatomic ions with 2+ charges, leading to similar chemical properties within this family.
c) 9, 17, 35, 53, 85. Halogens are the Group 7A/17 elements.
d) Z = 7 (N) < Z = 6 (C) < Z = 14 (Si) < Z = 32 (Ge). Remember that 'in order of increasing size" means smallest to largest, with the smallest listed first.
e) 34, 35, 36 (As, Z = 33, is a metalloid).

Chapter 11

Chemical Bonding

EXAMPLE 11.1-WB

In textbook Example 11.1, you found that the calcium and chloride ions are isoelectronic with an atom of argon. Identify one more cation and at least one more anion that are isoelectronic with an atom of argon.

K^+, Sc^{3+}, S^{2-}, and P^{3-}

If K loses one electron to become K^+, it is isoelectronic with argon. If S gains two electrons and P gains three electrons, each has the same electron configuration as argon. Note that the formations of K^+, S^{2-}, and P^{3-} are identical to the formations of Na^+, O^{2-}, and N^{3-}, respectively, the ions immediately above them in the periodic table. Unless you referred to Figure 11.1, you probably did not include the scandium ion, Sc^{3+}, in your answer to the question. From the electron configuration of scandium (Z = 21), $1s^22s^22p^63s^23p^64s^23d^1$, you might have guessed the charge on the ion would be 3+ because removal of the three highest-energy electrons leaves the configuration of the noble gas argon. Your guess would have been correct.

☑ TARGET CHECK 11.1-WB

Write the formulas and the electron configurations of a monatomic anion and a monatomic cation that are each isoelectronic with neon.

☑ TARGET CHECK 11.2-WB

Use Lewis diagrams to show the electron transfer that occurs when a sodium atom reacts with a bromine atom to form a sodium bromide crystal.

☑ TARGET CHECK 11.3-WB

a) Explain how half-filled electron orbitals overlap in the formation of a covalent bond between two atoms of iodine.

b) Identify the bonding electron pairs and the lone pairs in a molecule of butanoic acid, shown below:

```
       H    H    H  :O:
       |    |    |   ||   ..
  H — C — C — C — C — O — H
       |    |    |        ..
       H    H    H
```

EXAMPLE 11.2-WB

Arrange the following bonds in order of decreasing polarity, and place a δ^+ at the positive end of the bond and a δ^- at its negative end.

S—Br C—O Se—I

δ^+C—Oδ^- δ^+S—Brδ^- δ^+Se—Iδ^-

The C—O bond, with an electronegativity difference of 3.5 – 2.5 = 1.0, is the most polar of the three bonds. This is followed by the S—Br bond (2.8 – 2.5 = 0.3) and Se—I (2.5 – 2.4 = 0.1).

☑ TARGET CHECK 11.4-WB

How many electrons are in a single bond? a double bond? a triple bond?

☑ TARGET CHECK 11.5-WB

Can a chlorine atom be bonded to two other atoms? Can a hydrogen atom be bonded to two other atoms? Explain each answer.

Name: ______________________ Date: ______________________

ID: ______________________ Section: ______________________

Questions, Exercises, and Problems

Section 11.1: Monatomic Ions with Noble-Gas Electron Configurations

1. Write the electron configuration of the third-period elements that form monatomic ions that are isoelectronic with a noble-gas atom. Also identify the noble gas atoms having those configurations.

2. Identify by symbol two positively-charged monatomic ions that are isoelectronic with argon.

3. Write the symbols of two ions that are isoelectronic with the barium ion.

Section 11.2: Ionic Bonds

4. Aluminum oxide is an ionic compound. Sketch the transfer of electrons from aluminum atoms to oxygen atoms that accounts for the chemical formula of the compound Al_2O_3.

5. When potassium and chlorine react to form an ionic compound, why is there only one chlorine atom for each potassium atom instead of two?

Name: ______________________ Date: ______________________

ID: ______________________ Section: ______________________

Section 11.3: Covalent Bonds

6. What is the meaning of *orbital overlap* in the formation of a covalent bond?

7. Sketch the formation of two covalent bonds by an atom of sulfur in making a molecule of hydrogen sulfide, H_2S.

8. Circle the lone electron pairs in the following Lewis diagram for hydrogen chloride:

$$\mathrm{H} - \underset{\cdot\cdot}{\overset{\cdot\cdot}{\mathrm{Cl}}}:$$

9. Show how atoms achieve the stability of noble-gas atoms in forming covalent bonds.

Section 11.4: Polar and Nonpolar Covalent Bonds

10. Compare the electron cloud formed by the bonding electron pair in a polar bond with that in a nonpolar bond.

Name: ______________________ Date: ______________________

ID: ______________________ Section: ______________________

11. Refer to Figure 11.7 and list the following bonds in order of decreasing polarity: S—O, N—Cl, C—C.

12. For each bond in Question 11, identify the positive pole, if any.

13. Identify the trends in electronegativities that may be observed in the periodic table.

Section 11.5: Multiple Bonds

14. Double bonds and triple bonds can conform to the octet rule. Could a quadruple bond (4) obey that rule? Why or why not?

Section 11.6: Atoms That Are Bonded to Two or More Other Atoms

15. An atom, X, is bonded to another atom by a double bond. What is the largest number of *additional atoms*—don't count the one to which it is already bonded— to which X may be bonded and still conform to the octet rule? What is the minimum number? Justify your answers.

16. A molecule contains a triple bond. Theoretically, what are the maximum and minimum numbers of atoms that can be in the molecule and still conform to the octet rule? Sketch Lewis diagrams that justify your answer.

Name: ______________________ Date: ______________________

ID: ______________________ Section: ______________________

Section 11.7: Exceptions to the Octet Rule

17. Because nitrogen has five valence electrons, it is sometimes difficult to fit a nitrogen atom into a Lewis diagram that obeys the octet rule. Why is this so? Without actually drawing them, can you tell which of the following species does not have a Lewis diagram that satisfies the octet rule: N_2O, NO_2, NF_3, NO, N_2O_3, N_2O_4, NOCl, NO_2Cl?

Section 11.8: Metallic Bonds

18. What are the meanings of the terms *localized* and *delocalized*, when used to describe electrons in a compound?

19. How would a particulate-level illustration differ if you were to draw a calcium crystal instead of a potassium crystal?

20. Are alloys pure substances or mixtures? Are they compounds? Explain your answers.

General Questions

21. Explain why the total energy of a system changes in the formation of (a) an ionic bond and (b) a covalent bond.

Name: ______________________ Date: ______________________

ID: ______________________ Section: ______________________

22. Considering bonds between the following pairs of elements, which are most apt to be ionic and which are most apt to be covalent: sodium and sulfur; fluorine and chlorine; oxygen and sulfur? Explain your choice in each case.

23. Identify the pairs among the following that are *not* isoelectronic: (a) Ne and Na^{+}; (b) S^{2-} and Cl^{-}; (c) Mg^{2+} and Ar, (d) K^{+} and S^{2-}; (e) Ba^{2+} and Te^{2-} (Te, Z = 52).

24. Is there any such thing as a completely nonpolar bond? If yes, give an example.

More Challenging Problems

25. The metallic bond is neither ionic nor covalent. Explain, according to the octet rule, why this is so.

26. How do the energy and stability of bonded atoms and noble-gas electron configurations appear to be related in forming covalent and ionic bonds?

Name: ______________________ Date: ______________________

ID: ______________________ Section: ______________________

27. Arrange the following bonds in order of increasing polarity: Na—O; Al—O; S—O; K—O; Ca—O. If any two bonds cannot be positively placed relative to each other, explain why. You may consult only a full periodic table when answering this question.

28. How is it possible for central atoms in molecules to be surrounded by five or six bonding electron pairs when there are only four valence electrons from the *s* and *p* orbitals of any atom?

29. Suggest a reason why BF_3 behaves as a molecular compound, whereas AlF_3 appears to be ionic.

Answers to Target Checks

1. The anions isoelectronic with neon are N^{3-}, O^{2-}, and F^-. The monatomic cations isoelectronic with neon are Na^+, Mg^{2+}, and Al^{3+}. All have the electron configuration $1s^2 2s^2 2p^6$.
2. The Lewis diagram of sodium is the chemical symbol of sodium, Na, with one valence electron dot, because sodium is in Group 1A/1. The Lewis diagram of bromine is the chemical symbol of bromine, Br, with seven valence electron dots, because bromine is in Group 7A/17. The one electron dot from the sodium atom will transfer to the bromine atom, giving the newly formed bromide ion, Br^-, eight electron dots and leaving none on the newly formed sodium ion, Na^+.
3. a) Each separated iodine atom has all of its orbitals filled with the exception of one of the $5p$ orbitals, which has only one electron. This is a half-filled electron orbital. When two of these half-filled orbitals overlap, a pair of electrons is shared between the two orbitals, forming a covalent, or electron-sharing, bond.
   b) The lone pairs are represented by electron dots on the oxygen atoms. All other electron pairs are bonding pairs, represented by dashes.
4. A single bond has two electrons; a double bond has four; a triple bond has six.
5. A chlorine atom has only one unpaired electron, so it would appear from the information in this chapter that it could form only one bond. In fact, a chlorine atom can form three additional bonds in which each of its lone pairs becomes a bonding pair to an atom with fewer than seven valence electrons. A hydrogen atom cannot bond to to other atoms. A bonded hydrogen atom has the two-electron configuration of the Group 8A/18 gas helium. It therefore only has one single bond.

Chapter 12

Structure and Shape

EXAMPLE 12.1-WB

Write the Lewis diagrams for the silicon tetrahydride and carbon tetrabromide molecules.

Start with the valence electron count and central atom determination for the silicon tetrahydride molecule.

SiH_4: 4 (Si) + 4 × 1 (H) = 8 valence electrons; Si is the central atom

Draw the tentative diagram and compare the number of electrons in it to the number available.

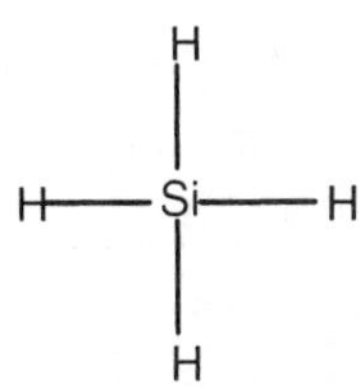

The tentative diagram has eight electrons, and eight are available. The silicon tetrahydride Lewis diagram is complete.

Carbon tetrabromide is next. Take it all the way.

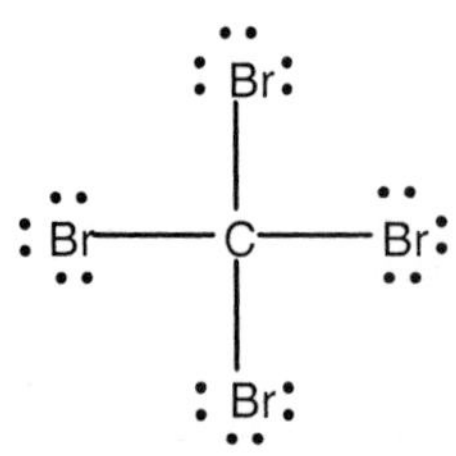

CBr_4: 4 (C) + 4 × 7 (Br) = 32 valence electrons. Carbon is less electronegative than bromine, so C is central. The number of electrons in the tentative diagram match the number of valence electrons.

In both molecules, the central atom is from Group 4A/14. Furthermore, each terminal atom, H and Br, has one valence electron to share to form a single bond with one of the four valence electrons on the central atom. Thus both molecules have four single bonds to the central atom.

EXAMPLE 12.2-WB

Draw Lewis diagrams for hydrogen bromide and hydrogen iodide.

Determine the number of valence electrons in each molecule, draw the tentative diagram, and compare the number of electrons in the tentative diagram to the number available.

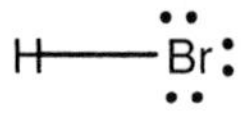

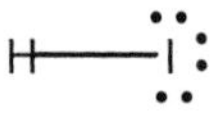

HBr and HI: 1 (H) + 7 (Br, I) = 8 valence electrons. Both tentative diagrams have 8 valence electrons.

Again, notice that since hydrogen is bonded to two different elements in the same group in the periodic table, the Lewis diagrams are similar, with only the elemental symbol of the Group 7A/17 differing.

EXAMPLE 12.3-WB

Draw the Lewis diagram for CH_2O.

How many valence electrons does this molecule have? Which atom will be central?

4 (C) + 2 × 1 (H) + 6 (O) = 12 valence electrons. Carbon is central.

Now draw the tentative diagram. Count the number of valence electrons on it, and compare that number with the number available.

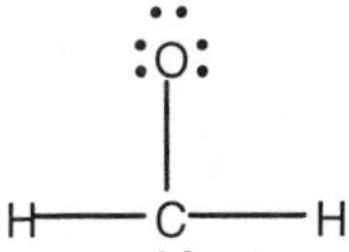

14 valence electrons; 2 too many

When you have too many valence electrons, cash in two pairs for one. Erase the lone pair on the central atom and one of the lone pairs on oxygen, and replace them with an additional bond between carbon and oxygen.

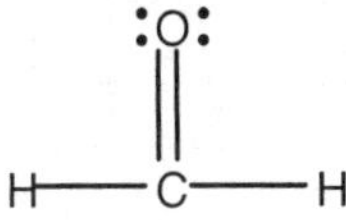

Is the octet rule satisfied for carbon and oxygen? Does each hydrogen atom have one single bond?

Yes, all atoms have the appropriate number of valence electrons. The diagram is complete.

EXAMPLE 12.4-WB

Write the Lewis diagram for the nitrite ion.

Begin with a valence electron count and central atom determination.

NO_2^-: 5 (N) + 2 × 6 (O) + 1 (charge) = 18 valence electrons; nitrogen is central.

Draw the tentative diagram and compare electron counts.

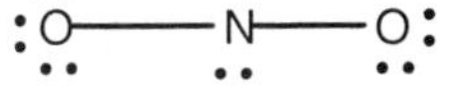

:Ö—N̈—Ö:

The tentative diagram has 20 valence electrons, 2 too many.

Make the needed correction and check to be sure each atom has an octet. Add the square brackets and superscript charge to your final diagram.

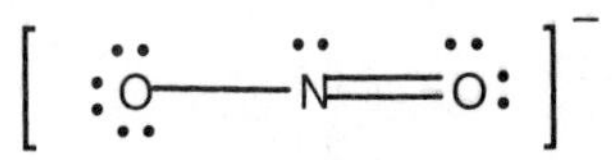

$$\left[\,:\ddot{\underset{\cdot\cdot}{O}}—\ddot{N}=\ddot{O}:\,\right]^-$$

You may have drawn the double bond to either oxygen atom; the diagrams are the same.

EXAMPLE 12.5-WB

Draw Lewis diagrams for BeH_2 and $AlCl_3$.

Both of these species are exceptions to the octet rule. Beryllium requires two single bonds to it and aluminum requires three. See if you can draw the diagram for each without futher guidance.

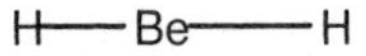

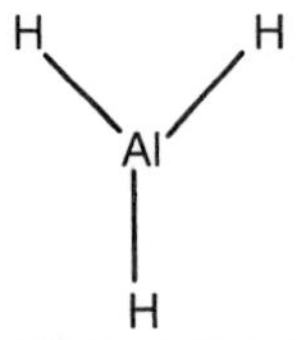

H—Be—H

H H
Al
|
H

EXAMPLE 12.6-WB

Write the Lewis diagram for butane, C_4H_{10}.

This is very similar to Example 12.6 in the textbook; this example has one more —CH_2— unit. Complete the problem.

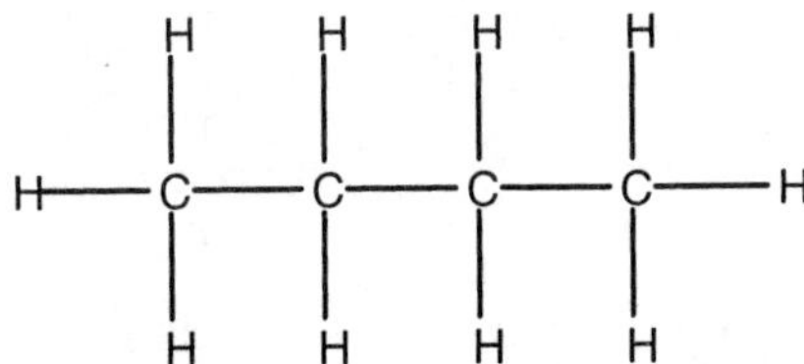

4×4 (C) + 10×1 (H) = 26 valence electrons
All valence electrons are accounted for in the 13 single bonds in the diagram.

EXAMPLE 12.7-WB

What is the Lewis diagram for nitrogen?

Start with the valence electron count and the tentative diagram. Count the number of valence electrons in your tentative diagram.

$2 \times 5 = 10$ valence electrons

:N̈—N̈:

The tentative diagram has 14 valence electrons.

To correct this discrepancy, remove a lone pair from each atom, subtracting four electrons, and add a bonding pair between the two atoms, adding back two electrons, for a net loss of two electrons. All the while, you will maintain the octet on each atom. Repeat the process to go from 12 valence electrons to 10.

:N≡N:

The nitrogen molecule has a triple bond between the two atoms.

EXAMPLE 12.8-WB

Draw two different Lewis diagrams for C_3H_8O.

One diagram has the oxygen atom between carbon atoms, and the other has all of the carbon atoms bonded to one another. Complete the example.

```
     H         H    H                  H    H    H
     |   ..    |    |                  |    |    |    ..
H ── C ── O ── C ── C ── H        H ── C ── C ── C ── O ── H
     |   ..    |    |                  |    |    |    ..
     H         H    H                  H    H    H
```

The oxygen atom may be placed between any two carbon atoms; the diagrams are equivalent. Likewise, the —O—H group can be at either end of the three-carbon chain.

EXAMPLE 12.9-WB

Determine the molecular geometry of $BeCl_2$.

Step 1 is to draw the Lewis diagram.

$:\ddot{Cl}—Be—\ddot{Cl}:$

Step 2 is to count the number of electron pairs around the central atom. Do that and give the name of the electron-pair geometry that corresponds to the number of electron pairs.

There are two electron pairs around the central atom. This yields a linear electron-pair geometry.

Step 3 is to determine the electron-pair (already completed) and molecular geometries. Write the name of the molecular geometry.

Linear

Two electron pairs, both bonded, yield a linear molecular geometry.

Step 4 is to sketch a three-dimensional representation of the molecule. All three atoms are in the same plane, and you know the bond angle for a linear molecule. Draw the sketch.

EXAMPLE 12.10-WB

Experimental evidence indicates that $GeCl_2$ is comprised of two chlorine atoms single bonded to the germanium (Z = 32) atom. The germanium atom has a lone pair. Its Lewis diagram is

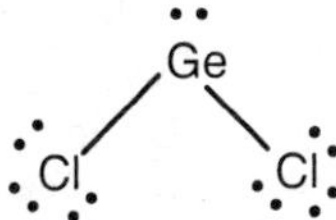

What is its molecular geometry? Sketch the molecule.

What is the electron-pair geometry of this molecule?

Trigonal planar

Three electron pairs around the central atom gives a trigonal planar electron-pair geometry.

What is the molecular geometry?

Angular

Three electron pairs, two bonded, yields an angular geometry.

Draw a sketch of the molecule.

EXAMPLE 12.11-WB

Sketch a three-dimensional ball-and-stick diagram of nitrogen trifluoride. What is the word descripition of this molecular geometry?

Begin with the valence electron count and the Lewis diagram.

5 (N) + 3 × 7 (F) = 26 valence electrons

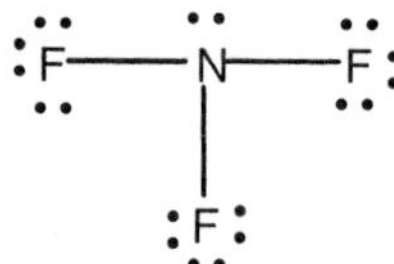

What are the electron-pair and molecular geometries?

Electron-pair geometry: tetrahedral; molecular geometry: trigonal pyramidal.

Draw a sketch of this molecule. If you are having trouble visualizing it, start with a sketch of a tetrahedral molecule and erase one of the bonded atoms.

EXAMPLE 12.12-WB

When sulfur dioxide emissions enter the atmosphere, they can react with oxygen and then water to form acid rain. Determine the molecular geometry of sulfur dioxide and draw a sketch of the molecule.

Begin with valence electron count and the tentative Lewis diagram.

6 (S) + 2 × 6 (O) = 18 valence electrons

:Ö—S̈—Ö:

The tentative diagram has 20 valence electrons.

Make the needed adjustment to the tentative diagram. How many regions of electron density surround the central atom?

O—S=O

There are three regions of electron density around the sulfur atom: the single bond, the double bond, and the unshared pair.

Write the names of the electron-pair geometry and the molecular geometry.

Electron-pair geometry: trigonal planar; molecular geometry: angular
Three regions of electron density yield a trigonal planar geometry, and two regions are bonded to an atom, yielding a angular molecular geometry.

Now draw a sketch that reflects this angular geometry.

EXAMPLE 12.13-WB

In Example 12.1-WB, you drew the Lewis diagram for carbon tetrabromide, CBr_4. Consider this molecule as well as $CHBr_3$, CH_2Br_2, CH_3Br, and CH_4. Which molecule(s) is/are polar? Which is/are nonpolar? Explain your reasoning in each case.

Start with the Lewis diagrams. Do all five.

CBr_4: 4 (C) + 4 × 7 (Br) = 32 valence electrons
$CHBr_3$: 4 (C) + 1 (H) + 3 × 7 (Br) = 26 valence electrons
CH_2Br_2: 4 (C) + 2 × 1 (H) + 2 × 7 (Br) = 20 valence electrons
CH_3Br: 4 (C) + 3 × 1 (H) + 7 (Br) = 14 valence electrons
CH_4:4 (C) + 4 × 1 (H) = 8 valence electrons

Now sketch a three-dimensional ball-and-stick diagram for each. Shade the bromine atom circles so that they may clearly be distinguished from the hydrogens.

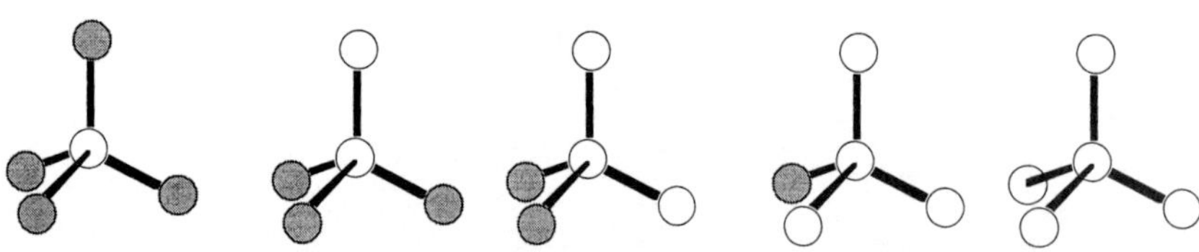

Now carefully consider the net effect of the polar bonds. Bromine is more electronegative than carbon, and hydrogen is slightly less electronegative than carbon. Consider your three-dimensional diagrams and indicate which is/are polar and which is/are not.

$CHBr_3$, CH_2Br_2, and CH_3Br are polar, CBr_4 and CH_4 are not.
In CBr_4 and CH_4, the polar bonds are arranged symmetrically, yielding a net nonpolar molecule. In the others, the polar bonds are not balanced, which gives a polar molecule.

Name: ______________________ Date: ______________________

ID: ______________________ Section: ______________________

Questions, Exercises, and Problems

Section 12.1: Drawing Lewis Diagrams

Write Lewis diagrams for each of the following sets of molecules.

1. HI, H_2O, NCl_3

2. CO_2, SF_2, BrO_3^-

3. BrO_-, $H_2PO_4^-$, ClO_4^-

4. $CHFCl_2$, CHF_3, $CClI_3$

Name: ______________________ Date: ______________________

ID: ______________________ Section: ______________________

There are two or more acceptable diagrams for most species in Questions 5 to 9.

5. $C_2H_2Cl_4$, $C_2H_2Cl_2F_2$, $C_3H_4Br_3I$

6. C_4H_8, C_2H_6O, $C_3H_8O_2$

7. C_6H_{14}, C_4H_6O, $C_2H_2F_2$

8. Butanoic acid, C_3H_7COOH

9. NO_2^+, N_2O, NO^+

Name: ______________________ Date: ______________________

ID: ______________________ Section: ______________________

Section 12.2: Electron-Pair Repulsion: Electron-Pair Geometry
Section 12.3: Molecular Geometry
Section 12.4: The Geometry of Multiple Bonds

Questions 10 to 15: For each molecule or ion, or for the atom specified in a molecule or ion, write the Lewis diagram, then describe (a) the electron-pair geometry and (b) the molecular geometry predicted by the electron-pair repulsion theory. Also sketch the three-dimensional ball-and-stick representation of each molecule or ion in Questions 10 and 11.

10. BCl_3, PH_3, H_2S

11. BrO^-, ClO_3^-, PO_4^{3-}

12. Each carbon atom in C_3H_7OH

Name: ____________________ Date: ____________________

ID: ____________________ Section: ____________________

13. Nitrogen atom in $C_2H_5NH_2$

14. Each carbon atom in C_2H_2

15. Carbon atom in HCN

16. The Lewis diagram of a certain compound has the element E as its central atom. The bonding and lone-pair electrons around E are shown. What is the molecular geometry around E?

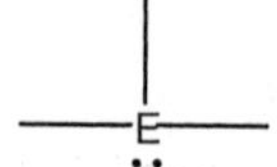

Name: ______________________________ Date: ______________________

ID: ______________________________ Section: ______________________

Questions 17 to 19: For each space-filling or ball-and-stick model shown, identify the electron-pair and molecular geometry. There are no "hidden" atoms in any of the models.

17. 

18. 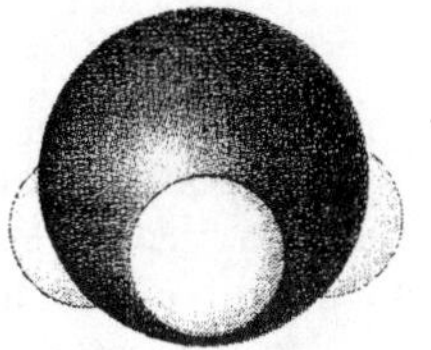

19. 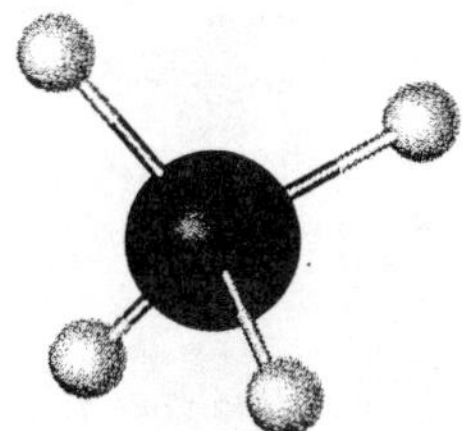

Section 12.5: Polarity of Molecules

20. Consider the following general Lewis diagrams

a) A—B—A (two lone pairs on B)

b) A—B—A

c) 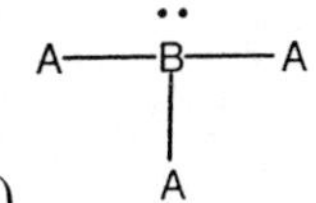

d) 

e) A—B—A (with A above and below B)

Which of the compounds have nonpolar molecules but polar bonds? Explain.

Name: ______________________ Date: ______________________

ID: ______________________ Section: ______________________

21. Is the carbon tetrachloride molecule, CCl_4, which contains four polar bonds (electronegativity difference 0.5) polar or nonpolar? Explain.

22. Describe the shapes and compare the polarities of HF and HBr molecules. In each case, identify the end of the molecule that is more positive.

23. Sketch the Lewis diagram of the water molecule, paying particular attention to the bond angle and using arrows to indicate the polarity of each bond. Then sketch the methanol molecule, $HOCH_3$, again using arrows to show bond polarity. Predict the approximate shape of both molecules around the oxygen atom. Also predict relative polarities of the two molecules and explain your prediction.

The answers for Questions 24 and 25 are to be selected from the group of Lewis diagrams that follows. Each question may have more than one answer.

a) $[BH_4]^-$ (H—B—H, with H above and below B)

b) :Cl—S—Cl: (with lone pairs)

c) :Cl—P—Cl: with :Cl: below P (with lone pairs)

d) $[NH_4]^+$ (H—N—H, with H above and below N)

e) :F—Be—F: (with lone pairs)

f) B bonded to three F atoms (:F: above, :F and F: below, with lone pairs)

24. Which species have tetrahedral shapes?

25. Which neutral molecules are polar?

Name: ______________________ Date: ______________________

ID: ______________________ Section: ______________________

Section 12.6: The Structure of Some Organic Compounds (Optional)

26. What distinguishes an organic compound from an inorganic compound?

27. Identify the hydrocarbons among the following: CH_3OH, $CH_3(CH_2)_6CH_3$, C_6H_6, $CH_2(NH_2)_2$, $C_{18}H_8$.

28. Distinguish between the *structure* of an organic molecule and the *shape* of the molecule.

29. Why can a zigzag chain of carbon atoms be drawn as a straight line in a Lewis diagram?

30. How are alcohols and ethers similar to water in structure and shape? What distinguishes between alcohols and ethers?

31. How do carboxylic acids differ from other acids, such as hydrochloric, sulfuric, and carbonic acid?

Name: ______________________ Date: ______________________

ID: ______________________ Section: ______________________

General Questions

32. One kind of C_6H_{12} molecule has its carbon atoms in a ring. Draw the Lewis diagram.

33. Draw Lewis diagrams for these five acids of bromine: HBr, HBrO, $HBrO_2$, $HBrO_3$, $HBrO_4$.

More Challenging Problems

34. Describe the shapes of C_2H_6 and C_2H_4. In doing so, explain why one molecule is planar and the other molecule cannot be planar.

35. Draw two different Lewis diagrams of C_4H_6.

Name: ______________________ Date: ______________________

ID: ______________________ Section: ______________________

36. Compare Lewis diagrams for CCl_4, SO_4^{2-}, ClO_4^{-}, and PO_4^{3-}. Identify two things that are alike about these diagrams and the way they are drawn. From these generalizations, can you predict the Lewis diagram of SeO_4^{2-} and CI_4?

Chapter 13

The Ideal Gas Law and Its Applications

☑ TARGET CHECK 13.1-WB

One of two identical containers holds oxygen, and the other container holds helium. Both gases exert a pressure of 1.19 atm at 21°C. Compare (a) the number of particles in each container, (b) the number of atoms in each container, and (c) the mass of the contents of each container. In each case, state whether one is greater than the other or if they are equal, and explain your reasoning. Give as much quantitative detail as possible in your explanations.

EXAMPLE 13.1-WB

At 23°C, a 655-mL sample of argon exerted a pressure of 839 torr. It was determined that the sample contained 0.0298 mole of the gas. Use these data to find the value of R in L • torr/mol • K.

Use Equation 13.3 to solve the problem. Be careful with the volume of the gas.

GIVEN: P = 839 torr; V = 655 mL (0.655 L); n = 0.0298 mol; T = 23°C (296 K)

WANTED: R in L • torr/mol • K

EQUATION: $R = \frac{PV}{nT} = \frac{839 \text{ torr} \times 0.655 \text{ L}}{0.0298 \text{ mol} \times 296 \text{ K}} = 62.4 \text{ L} \cdot \text{torr/mol} \cdot \text{K}$

EXAMPLE 13.2-WB

How many moles of neon are in a 5.00-L gas cylinder at 28°C if they exert a pressure of 8.65 atm?

Write the *GIVEN*, *WANTED*, and the *EQUATION* solved for the unknown variable.

GIVEN: V = 5.00 L; T = 28°C (301 K); P = 8.65 atm

WANTED: n *EQUATION:* $n = \frac{PV}{RT}$

Substitute the known values and calculate the final answer.

$$n = \frac{PV}{RT} = 8.65 \text{ atm} \times 5.00 \text{ L} \times \frac{\text{mol} \cdot \text{K}}{0.0821 \text{ L} \cdot \text{atm}} \times \frac{1}{301 \text{ K}} = 1.75 \text{ mol}$$

☑ TARGET CHECK 13.2-WB

At what temperature (°C) will the pressure of 0.0743 mole of helium in a 2.16-L container be 682 torr?

EXAMPLE 13.3-WB

What is the volume of 0.787 grams of carbon dioxide at 1.07 atm and 15°C?

There are three straightforward givens in the problem statement and two "hidden" givens. Try to identify all five.

GIVEN: 0.787 g; 1.07 atm; 15°C (288 K); MM = 44.01 g/mol; R = 0.0821 L • atm/mol • K
One of the hidden givens is the molar mass of CO_2, (12.01 + 2 × 16.00) g/mol = 44.01 g/mol. The other is the ideal gas constant with pressure in atm.

You can now use the PV = mRT/MM form of the ideal gas equation to complete the problem.

$$V = \frac{mRT}{P(MM)} = 0.787\ g \times \frac{0.0821\ L \cdot atm}{mol \cdot K} \times 288\ K \times \frac{1}{1.07\ atm} \times \frac{mol}{44.01\ g} = 0.395\ L$$

EXAMPLE 13.4-WB

Calculate the mass of nitrogen, oxygen, water vapor, carbon dioxide, some noble gases, and other gases in a 12 inch × 12 inch × 12 inch box of air if it exerts a pressure of 0.897 atm at 18°C. Assume 29 g/mol as the average molar mass of the molecules in the box.

We'll spare you one part of the calculation: the volume of the one-cubic-foot box is 28 liters. Now solve Equation 13.6 for mass, substitute your numbers, and discover something new about this sea of gas that we live in.

GIVEN: 28 L; 0.897 atm; 18°C (291 K); 29 g/mol *WANTED:* mass (assume g)

EQUATION: $m = \frac{PV(MM)}{RT} = 0.897\ atm \times 28\ L \times \frac{29\ g}{mol} \times \frac{mol \cdot K}{0.0821\ L \cdot atm} \times \frac{1}{291\ K} = 3.0 \times 10^1\ g$

A cubic foot of air weighs thirty grams, which is the mass of about 2 tablespoons of water.

EXAMPLE 13.5-WB

Which gas is more dense at 17°C and 0.945 atm, C_2H_6 or NO? Calculate both densities to determine your answer.

There are two density calculations requested in the problem. Complete them, and then answer the question.

GIVEN: 17°C (290 K); 0.945 atm; 30.07 g/mol C_2H_6; 0.0821 L • atm/mol • K
WANTED: D (m/V)

EQUATION: $D = \frac{m}{V} = \frac{(MM)P}{RT} = \frac{30.07\text{ g}}{\text{mol}} \times 0.945\text{ atm} \times \frac{\text{mol} \cdot \text{K}}{0.0821\text{ L} \cdot \text{atm}} \times \frac{1}{290\text{ K}} = 1.19\text{ g/L}$

GIVEN: 17°C (290 K); 0.945 atm; 30.01 g/mol NO; 0.0821 L • atm/mol • K
WANTED: D (m/V)

EQUATION: $D = \frac{m}{V} = \frac{(MM)P}{RT} = \frac{30.01\text{ g}}{\text{mol}} \times 0.945\text{ atm} \times \frac{\text{mol} \cdot \text{K}}{0.0821\text{ L} \cdot \text{atm}} \times \frac{1}{290\text{ K}} = 1.19\text{ g/L}$

To three significant figures, the densities are the same.

The density of a gas at a given temperature and pressure is proportional to its molar mass.

EXAMPLE 13.6-WB

A 0.278-L sample of gas, measured at 125°C and 0.959 atm, is found to have a mass of 0.381 g. What is the molar mass of the gas?

Complete the problem.

GIVEN: 0.278 L; 15°C (398 K); 0.959 atm; 0.381 g *WANTED:* MM (g/mol)

EQUATION: $MM = \frac{mRT}{PV} = 0.381\text{ g} \times \frac{0.0821\text{ L} \cdot \text{atm}}{\text{mol} \cdot \text{K}} \times 398\text{ K} \times \frac{1}{0.959\text{ atm}} \times \frac{1}{0.278\text{ L}} = 46.7\text{ g/mol}$

EXAMPLE 13.7-WB

Find the molar volume of H_2S at 11°C and 845 torr.

Set up and solve the problem.

GIVEN: 11°C (284 K); 845 torr *WANTED:* MV (L/mol)

EQUATION: $MV \equiv \frac{V}{n} = \frac{RT}{P} = \frac{62.4\text{ L} \cdot \text{torr}}{\text{mol} \cdot \text{K}} \times 284\text{ K} \times \frac{1}{845\text{ torr}} = 21.0\text{ L/mol}$

EXAMPLE 13.8-WB

A gas sample occupies 7.61 L at a temperature and pressure at which the molar volume is 64.8 L/mol. How many moles are in the sample?

Complete the problem.

GIVEN: 7.61 L; 64.8 L/mol *WANTED:* mol

PER/PATH: $\text{L} \xrightarrow{64.8\ \text{L/mol}} \text{mol}$

$$7.61\ \text{L} \times \frac{1\ \text{mol}}{64.8\ \text{L}} = 0.117\ \text{mol}$$

EXAMPLE 13.9-WB

How many grams of sodium hydroxide can be recovered as a by-product of the chemical change that produces 71.9 liters of chlorine, measured at STP, by the reaction 2 NaCl(aq) + 2 $H_2O(\ell) \rightarrow$ 2 NaOH(aq) + $Cl_2(g)$ + $H_2(g)$?

Start by listing the *GIVEN* and *WANTED* for this problem.

GIVEN: 71.9 L Cl_2 *WANTED:* g NaOH

Now write out a *PER/PATH* that takes you from liters of chlorine to grams of sodium hydroxide.

PER/PATH: $\text{L Cl}_2 \xrightarrow{22.4\ \text{L Cl}_2/\text{mol Cl}_2} \text{mol Cl}_2$

$\xrightarrow{2\ \text{mol NaOH}/1\ \text{mol Cl}_2} \text{mol NaOH} \xrightarrow{40.00\ \text{g NaOH/mol NaOH}} \text{g NaOH}$

Complete the problem with the setup and final answer.

$$71.9 \text{ L } Cl_2 \times \frac{1 \text{ mol } Cl_2}{22.4 \text{ L } Cl_2} \times \frac{2 \text{ mol NaOH}}{1 \text{ mol } Cl_2} \times \frac{40.00 \text{ g NaOH}}{\text{mol NaOH}} = 257 \text{ g NaOH}$$

EXAMPLE 13.10-WB

How many grams of sodium hydroxide can be recovered as a by-product of the reaction that produces 71.9 liters of chlorine, measured at 1.17 atm and 38°C, by the reaction 2 NaCl(aq) + 2 $H_2O(\ell)$ → 2 NaOH(aq) + Cl_2(g) + H_2(g)?

Begin by writing the *GIVEN*, *WANTED*, and *PER/PATH* for this problem. You will need the molar volume for one of your *PER* expressions.

EQUATION: $MV \equiv \frac{V}{n} = \frac{RT}{P} = \frac{0.0821 \text{ L} \cdot \text{atm}}{\text{mol} \cdot \text{K}} \times (38 + 273) \text{ K} \times \frac{1}{1.17 \text{ atm}} = 21.8 \text{ L/mol}$

GIVEN: 71.9 L Cl_2 *WANTED:* g NaOH

PER/PATH: L Cl_2 $\xrightarrow{21.8 \text{ L/mol}}$ mol Cl_2 $\xrightarrow{2 \text{ mol NaOH/1 mol } Cl_2}$ mol NaOH $\xrightarrow{40.00 \text{ g NaOH/mol NaOH}}$ g NaOH

You are now ready to setup and solve the problem.

$$71.9 \text{ L } Cl_2 \times \frac{1 \text{ mol } Cl_2}{21.8 \text{ L } Cl_2} \times \frac{2 \text{ mol NaOH}}{1 \text{ mol } Cl_2} \times \frac{40.00 \text{ g NaOH}}{\text{mol NaOH}} = 264 \text{ g NaOH}$$

EXAMPLE 13.11-WB

Calculate the mass of C_4H_{10} that reacts if 52.1 L O_2, measured at 0.212 atm and 24°C, is used in burning in the reaction $2\ C_4H_{10}(g) + 13\ O_2(g) \rightarrow 8\ CO_2(g) + 10\ H_2O(\ell)$.

Take this one all the way.

GIVEN: 52.1 L O_2, 0.212 atm, 24°C (297 K) *WANTED:* g C_4H_{10}

EQUATION: $MV \equiv \frac{V}{n} = \frac{RT}{P} = \frac{0.0821\ L \cdot atm}{mol \cdot K} \times 297\ K \times \frac{1}{0.212\ atm} = 115\ L/mol$

$$52.1\ L\ O_2 \times \frac{1\ mol\ O_2}{115\ L\ O_2} \times \frac{2\ mol\ C_4H_{10}}{13\ mol\ O_2} \times \frac{58.12\ g\ C_4H_{10}}{mol\ C_4H_{10}} = 4.05\ g\ C_4H_{10}$$ *or*

$$52.1\ L\ O_2 \times \frac{mol \cdot K}{0.0821\ L \cdot atm} \times \frac{0.212\ atm}{297\ K} \times \frac{2\ mol\ C_4H_{10}}{13\ mol\ O_2} \times \frac{58.12\ g\ C_4H_{10}}{mol\ C_4H_{10}} = 4.05\ g\ C_4H_{10}$$

In this problem, the given quantity, rather than the wanted quantity, is a gas. That means that, in the alternative solution, you must divide by molar volume, RT/P, or multiply by its inverse, P/RT.

EXAMPLE 13.12-WB

How many grams of sodium hydroxide can be recovered as a by-product of the reaction that produces 71.9 liters of chlorine, measured at 1.17 atm and 38°C, by the reaction $2\ NaCl(aq) + 2\ H_2O(\ell) \rightarrow$ $2\ NaOH(aq) + Cl_2(g) + H_2(g)$?

Use the ideal gas equation to find the number of moles of chlorine.

GIVEN: 71.9 L; 1.17 atm; 311 K *WANTED:* mol Cl_2

EQUATION: $n = \frac{PV}{RT} = 1.17\ atm \times 71.9\ L \times \frac{mol \cdot K}{0.0821\ L \cdot atm} \times \frac{1}{311\ K} = 3.29\ mol$

Complete the problem.

GIVEN: 3.29 mol Cl_2 *WANTED:* g NaOH

PER/PATH: mol Cl_2 $\xrightarrow{\text{2 mol NaOH/1 mol Cl}_2}$ mol NaOH $\xrightarrow{\text{40.00 g NaOH/mol NaOH}}$ g NaOH

$$3.29 \text{ mol Cl}_2 \times \frac{2 \text{ mol NaOH}}{1 \text{ mol Cl}_2} \times \frac{40.00 \text{ g NaOH}}{\text{mol NaOH}} = 263 \text{ g NaOH}$$

EXAMPLE 13.13-WB

Calculate the mass of C_4H_{10} that reacts if 52.1 L O_2, measured at 0.212 atm and 24°C, is used in burning in the reaction $2\ C_4H_{10}(g) + 13\ O_2(g) \rightarrow 8\ CO_2(g) + 10\ H_2O(\ell)$.

This time the gas volume is given. Use the ideal gas equation to find the number of moles of oxygen in 52.1 liters at 0.212 atm and 24°C.

GIVEN: 52.1 L O_2, 0.212 atm, 24°C (297 K) *WANTED:* mol O_2

EQUATION: $n = \frac{PV}{RT} = 0.212 \text{ atm} \times 52.1 \text{ L} \times \frac{\text{mol} \cdot \text{K}}{0.0821 \text{ L} \cdot \text{atm}} \times \frac{1}{297 \text{ K}} = 0.453 \text{ mol O}_2$

Now use Steps 2 and 3 of the stoichiometry path to find g C_4H_{10}.

GIVEN: 0.453 mol O_2 *WANTED:* mass C_4H_{10} (assume g)

PER/PATH: mol O_2 $\xrightarrow{2 \text{ mol } C_4H_{10}/13 \text{ mol } O_2}$ mol C_4H_{10} $\xrightarrow{58.12 \text{ g } C_4H_{10}/\text{mol } C_4H_{10}}$ g C_4H_{10}

$$0.453 \text{ mol } O_2 \times \frac{2 \text{ mol } C_4H_{10}}{13 \text{ mol } O_2} \times \frac{58.12 \text{ g } C_4H_{10}}{\text{mol } C_4H_{10}} = 4.05 \text{ g } C_4H_{10}$$

A single setup, using the inverse of the molar volume, P/RT, gives

$$52.1 \text{ L} \times \frac{\text{mol} \cdot \text{K}}{0.0821 \text{ L} \cdot \text{atm}} \times \frac{0.212 \text{ atm}}{297 \text{ K}} \times \frac{2 \text{ mol } C_4H_{10}}{13 \text{ mol } O_2} \times \frac{58.12 \text{ g } C_4H_{10}}{\text{mol } C_4H_{10}} = 4.05 \text{ g } C_4H_{10}$$

EXAMPLE 13.14-WB

A total of 3.3 L of gaseous acetylene, C_2H_2, is burned in a welding operation. If the temperature of the flame is 2955°C and the gases burn at 0.96 atm pressure, what volume of oxygen is consumed in the combustion reaction? Assume that when the oxygen reacts, it is at the same temperature and pressure as the acetylene.

Begin by writing and balancing the reaction equation.

$$2\, C_2H_2 + 5\, O_2 \rightarrow 4\, CO_2 + 2\, H_2O$$

Determine the volume of oxygen that reacts.

GIVEN: 3.3 L C_2H_2 *WANTED:* volume O_2 (assume L)

PER/PATH: L C_2H_2 $\xrightarrow{5 \text{ L } O_2/2 \text{ L } C_2H_2}$ L O_2

$$3.3 \text{ L } C_2H_2 \times \frac{5 \text{ L } O_2}{2 \text{ L } C_2H_2} = 8.3 \text{ L } O_2$$

EXAMPLE 13.15-WB

What maximum volume of NOCl(g), measured at 35°C and 0.943 atm, can be obtained from the reaction of 5.17 L Cl_2(g) at 18°C and 49.0 atm? The equation is 2 NO(g) + Cl_2(g) → 2 NOCl(g).

Start with a table of initial and final values.

	Volume	Temperature	Pressure
Initial Value (1)	5.17 L	18°C; 291 K	49.0 atm
Final Value (2)	V_2	35°C; 308 K	0.943 atm

Determine the volume occupied by chlorine gas at the final conditions.

$$V_2 = V_1 \times \frac{P_1}{P_2} \times \frac{T_2}{T_1} = 5.17\ \text{L} \times \frac{49.0\ \text{atm}}{0.943\ \text{atm}} \times \frac{308\ \text{K}}{291\ \text{K}} = 284\ \text{L}$$

You can now complete the problem by setting up and solving the one-step conversion to volume of NOCl(g).

GIVEN: 284 L Cl_2 *WANTED:* volume NOCl (assume L)

PER/PATH: L Cl_2 $\xrightarrow{2\ \text{L NOCl}/1\ \text{L Cl}_2}$ L NOCl

$$284\ \text{L Cl}_2 \times \frac{2\ \text{L NOCl}}{1\ \text{L Cl}_2} = 568\ \text{L NOCl}$$

Name: ______________________ Date: ______________________

ID: ______________________ Section: ______________________

Questions, Exercises, and Problems

Section 13.2: The Volume–Amount (Avogadro's) Law

1. Compare the volumes of 1×10^{23} hydrogen molecules, 1×10^{23} oxygen molecules, and 2×10^{23} nitrogen molecules, all at the same temperature and pressure.

Section 13.4: The Ideal Gas Equation: Determination of a Single Variable

2. Find the pressure in torr produced by 0.0888 mole of carbon dioxide in a 5.00-liter vessel at 36°C.

Answer: [ ]

3. The pressure caused by 6.04 mol of nitrogen monoxide at a temperature of 18°C is 17.2 atm. What is the volume of the gas in liters?

Answer: [ ]

Name: ______________________ Date: ______________________

ID: ______________________ Section: ______________________

4. A 784-mL hydrogen lecture bottle is left with the valve slightly open. Assuming no air has mixed with the hydrogen, how many moles of hydrogen are left in the bottle after the pressure has become equal to an atmospheric pressure of 752 torr at a temperature of 22°C?

Answer: [ ]

5. At what temperature (°C) will 0.810 mol of chlorine in a 15.7-L vessel exert a pressure of 756 torr?

Answer: [ ]

6. How many moles of carbon monoxide must be placed into a 40.0-L tank to develop a pressure of 965 torr at 18°C?

Answer: [ ]

Name: ______________________ Date: ______________________

ID: ______________________ Section: ______________________

7. Find the volume of 0.621 mole of helium at –32°C and 0.771 atm.

Answer: [ ]

Section 13.5: Gas Density and Molar Volume

8. The STP density of an unknown gas is found to be 2.32 g/L. What is the molar mass of the gas?

Answer: [ ]

9. In Example 13.4-WB you used 29 g/mol as the "effective" molar mass of air. Use this value to calculate the density of air (a) at STP and (b) at 20°C and 751 torr.

Answer: [ ]

Name: ______________________ Date: ______________________

ID: ______________________ Section: ______________________

10. If the density of an unknown gas at 41°C and 2.61 atm is 1.61 g/L, what is its molar mass?

Answer: [ ]

11. The gas in an 8.07-liter cylinder at 13°C has a mass of 33.5 grams and exerts a pressure of 3.25 atm. Find the molar mass of the gas.

Answer: [ ]

12. NO_2 and N_2O_4 both have the same empirical (simplest) formula. At a temperature and pressure at which both substances are gases, can you tell without calculating which gas is more dense? Explain.

Name: ______________________ Date: ______________________

ID: ______________________ Section: ______________________

13. Compare the molar volumes of helium and neon at 30°C and 1.10 torr. What are their values in L/mol?

Answer: [ ]

14. Find the molar volume of acetylene, C_2H_2, at 21°C and 0.908 atm.

Answer: [ ]

Section 13.6: Gas Stoichiometry at Standard Temperature and Pressure

15. One small-scale laboratory method for preparing oxygen is to heat potassium chlorate in the presence of a catalyst: $2\ KClO_3(s) \rightarrow 2\ KCl(s) + 3\ O_2(g)$. Find the STP volume of oxygen that can be produced by 5.74 g $KClO_3$.

Answer: [ ]

Name: ______________________ Date: ______________________

ID: ______________________ Section: ______________________

16. The reaction used to produce chlorine in the laboratory is $2\ KMnO_4(aq) + 16\ HCl(aq) \rightarrow 2\ MnCl_2(aq) + 5\ Cl_2(aq) + 8\ H_2O(aq) + 2\ KCl(aq)$. Calculate the number of grams of potassium permanganate, $KMnO_4$, that are needed to produce 9.81 L of chlorine, measured at STP.

Answer: [ ]

Section 13.7: Gas Stoichiometry: Molar Volume Method
Section 13.8: Gas Stoichiometry: Ideal Gas Equation Method

Questions 17–20 may be solved by the molar volume method (Section 13.7) or by the ideal gas equation method (Section 13.8). Setup your work according to the section you studied.

17. One source of sulfur dioxide used in making sulfuric acid comes from sulfide ores by the reaction $4\ FeS_2(s) + 11\ O_2(g) \rightarrow 2\ Fe_2O_3(s) + 8\ SO_2(g)$. How many liters of SO_2, measured at 983 torr and 214°C, are produced by the reaction of 598 g FeS_2?

Answer: [ ]

18. How many grams of water must decompose by electrolysis to produce 23.9 L H_2, measured at 28°C and 728 torr?

Answer: [ ]

Name: ____________________ Date: ____________________

ID: ____________________ Section: ____________________

19. The reaction chamber in a modified Haber process for making ammonia by the direct combination of its elements is operated at 575°C and 248 atm. How many liters of nitrogen, measured at these conditions, will react to produce 9.16×10^3 grams of ammonia?

Answer: ____________________

20. When properly detonated, ammonium nitrate explodes violently, releasing hot gases: $NH_4NO_3(s) \rightarrow N_2O(g) + 2\ H_2O(g)$. If the total volume of gas produced, both dinitrogen and steam, is 82.3 L at 447°C and 896 torr, how many grams of NH_4NO_3 exploded?

Answer: ____________________

Section 13.9: Volume–Volume Gas Stoichiometry

21. Sulfur burns to SO_2 with a beautiful deep blue–purple flame, but with a foul, suffocating odor: $2\ S + O_2 \rightarrow 2\ SO_2$. (a) How many liters of O_2 are needed to form 35.2 L SO_2, both gases being measured at 741 torr and 26°C? (b) What if only the SO_2 is at those conditions, but the O_2 is at 17°C and 847 torr?

Answer: ____________________

Name: ______________________ Date: ______________________

ID: ______________________ Section: ______________________

22. Gaseous chlorine dioxide, ClO_2, is used to bleach flour and in water treatment. It is produced by the reaction of chlorine with sodium chlorite: $Cl_2 + 2\ NaClO_2 \rightarrow 2\ ClO_2 + 2\ NaCl$. How many liters of ClO_2, measured at 0.961 atm and 31°C, will be produced by 283 L Cl_2 at 2.91 atm and 21°C?

Answer: [ ]

23. In the natural oxidation of hydrogen sulfide released by decaying organic matter, the following reaction occurs: $2\ H_2S + 3\ O_2 \rightarrow 2\ SO_2 + 2\ H_2O$. How many liters of hydrogen sulfide, measured at 19°C and 549 torr, will be used in a reaction that also uses 704 mL O_2 at 159 torr and 26°C?

Answer: [ ]

General Questions

24. Find the mass of 57.9 liters of krypton at 775 torr and 6°C.

Answer: [ ]

Name: ______________________ Date: ______________________

ID: ______________________ Section: ______________________

25. At 17°C and 0.835 atm, 16.2 liters of ammonia has a mass of 9.68 g. What is the molar volume of ammonia at those conditions? (This is easier than it may seem!)

Answer: [ ]

26. A 7.60-g sample of pure liquid is vaporized at 183°C and 179 torr. At these conditions it occupies 3.87 L. What is the molar mass of the substance?

Answer: [ ]

More Challenging Problems

27. The density of nitrogen at 0.913 atm and 18°C is 1.07 g/L. Explain how this shows that the formula of nitrogen is N_2 rather than just N.

Answer: [ ]

Name: ______________________ Date: ______________________

ID: ______________________ Section: ______________________

28. An organic chemist has produced a solid she believes to be pure; she expects a molar mass of 346 g/mol. Using 4.08 grams of the solid, she melts and then boils it in a 3.36-L vacuum chamber at 117 torr and 243°C. She is disappointed; the molar mass is close to what she expected, but not close enough. (a) What molar mass did she find? (b) Her finding suggested to her what her mistake might have been. Does it suggest anything to you? [Part (b) is beyond the scope of Chapter 13, but you might have an inspiration if you have studied Chapter 12.)

Answers to Target Checks

1. (a) The number of particles in each container is equal. Avogadro's Law states that at constant temperature and pressure, containers of equal volume contain equal numbers of particles. The number of oxygen molecules in one container equals the number of helium atoms in the other. (b) The number of atoms in the oxygen container is twice the number of atoms in the helium container because the oxygen molecule is made up of two oxygen atoms. The number of oxygen *molecules* is equal to the number of helium atoms. (c) The mass in the oxygen container is 32 ÷ 4 = 8 times the mass in the helium container (to the ones place, the molar mass of oxygen is 32 g/mol, and the molar mass of helium is 4 g/mol).

2. $$T = \frac{PV}{nR} = 682 \text{ torr} \times 2.16 \text{ L} \times \frac{1}{0.0743 \text{ mol}} \times \frac{\text{mol} \cdot \text{K}}{62.4 \text{ L} \cdot \text{torr}} = 318 \text{ K}; 318 - 273 = 45°\text{C}$$

Chapter 14

Combined Gas Law Applications

EXAMPLE 14.1-WB

A container holds 2.7×10^{-2} mole of helium at STP. What is the volume of the container, expressed in milliliters?

Write the *GIVEN, WANTED,* and *PER/PATH* for the problem.

GIVEN: 2.7×10^{-2} mol He *WANTED:* mL He

PER/PATH: mol He $\xrightarrow{\text{22.4 L He/mol He}}$ L He $\xrightarrow{\text{1000 mL He/L He}}$ mL He

Now you are ready to write the setup and calculate the answer.

$$2.7 \times 10^{-2} \text{ mol He} \times \frac{22.4 \text{ L He}}{\text{mol He}} \times \frac{1000 \text{ mL He}}{\text{L He}} = 6.0 \times 10^{2} \text{ mL He}$$

EXAMPLE 14.2-WB

What is the volume of 0.744 mole of chlorine at 1.36 atm and –7°C?

In textbook Example 14.2, you determined that the molar volume of any gas at 1.36 atm and –7°C is 16.0 L/mol. Don't hesitate to use that information in the solution to this problem. Go all the way to the final answer.

GIVEN: 0.744 mol Cl_2; 16.0 L Cl_2/mol Cl_2 *WANTED:* volume Cl_2 (assume L)

PER/PATH: mol Cl_2 $\xrightarrow{16.0 \text{ L } Cl_2/\text{mol } Cl_2}$ mol Cl_2

$$0.744 \text{ mol } Cl_2 \times \frac{16.0 \text{ L } Cl_2}{\text{mol } Cl_2} = 11.9 \text{ L } Cl_2$$

☑ TARGET CHECK 14.1-WB

(a) Determine the molar volume of nitrogen at 693 torr and 53°C. (b) Use the molar volume to find the number of moles of nitrogen in a 10.0-L container at that pressure and temperature.

EXAMPLE 14.3-WB

Calculate the volume occupied by 3.95 grams of neon at a temperature and pressure at which its density is 0.865 g/L.

Complete the solution.

GIVEN: 3.95 g Ne; 0.865 g Ne/L Ne *WANTED:* volume Ne (assume L)

PER/PATH: g Ne $\xrightarrow{0.865 \text{ g Ne/L Ne}}$ L Ne

$$3.95 \text{ g Ne} \times \frac{1 \text{ L Ne}}{0.865 \text{ g Ne}} = 4.57 \text{ L Ne}$$

EXAMPLE 14.4-WB

Calculate the density of carbon dioxide at STP. What is the weight in pounds of 1.50 liters of carbon dioxide at STP? 1 lb = 453.6 g

Begin with the calculation of the STP density of carbon dioxide.

GIVEN: 22.4 L CO_2/mol CO_2; 44.01 g CO_2/mol CO_2 *WANTED:* Density (assume g/L)

EQUATION: $D \equiv \frac{m}{V} = \frac{44.01 \text{ g } CO_2/\text{mol } CO_2}{22.4 \text{ L } CO_2/\text{mol } CO_2} = 1.96 \text{ g/L}$

PLAN the second part of the solution.

GIVEN: 1.50 L CO_2 *WANTED:* lb CO_2

PER/PATH: L CO_2 $\xrightarrow{1.96 \text{ g } CO_2/\text{L } CO_2}$ g CO_2 $\xrightarrow{1 \text{ lb } CO_2/453.6 \text{ g } CO_2}$ lb CO_2

Finish the problem with the setup and final answer.

$$1.50 \text{ L } CO_2 \times \frac{1.96 \text{ g } CO_2}{\text{L } CO_2} \times \frac{1 \text{ lb } CO_2}{453.6 \text{ g } CO_2} = 6.48 \times 10^{-3} \text{ lb } CO_2$$

EXAMPLE 14.5-WB

The density of an unknown gas is 1.16 g/L at STP. Find the molar mass of the gas.

One ratio is given in the problem statement. What is the other ratio needed to answer this question?

22.4 L/mol

The molar volume of a gas at STP is 22.4 L/mol.

The ratios g/L and L/mol can be combined to find the desired ratio, g/mol. Calculate the answer to the question.

$$\frac{1.16\text{ g}}{\text{L}} \times \frac{22.4\text{ L}}{\text{mol}} = 26.0\text{ g/mol}$$

EXAMPLE 14.6-WB

(a) Find the density of hydrogen at 0.897 atm and 18°C. (b) Find the molar volume of dihydrogen sulfide (often called hydrogen sulfide) when its density is 1.18 g/L.

These two problems are typical three-ratio problems at a given temperature and pressure. Consider each separately, starting with Part (a). What is the definition of density?

$$D \equiv \frac{m}{V}$$

The general strategy in a three-ratio problem is to combine two ratios to form the third. To find density, you need one ratio with mass in it and another ratio that has volume. Other than density, what ratio do you know that has mass in its numerator or denominator? What ratio has volume in it?

$$MM \equiv \frac{g}{mol} \quad \text{and} \quad MV \equiv \frac{L}{mol}$$

You need to calculate each of these ratios. Start with molar mass. What is the molar mass of hydrogen?

$2(1.008\text{ g/mol H}) = 2.016\text{ g/mol }H_2$

Now consider molar volume. You know the molar volume of a gas at STP, and you need the molar volume at 0.897 atm and 18°C. Setup and solve.

	Molar Volume	Temperature	Pressure
Initial Value (1)	22.4 L/mol	0°C;273 K	1 atm
Final Value (2)	MV_2	18°C; 291 K	0.897 atm

$$MV_2 = MV_1 \times \frac{P_1}{P_2} \times \frac{T_2}{T_1} = 22.4 \text{ L/mol} \times \frac{1 \text{ atm}}{0.897 \text{ atm}} \times \frac{291 \text{ K}}{273 \text{ K}} = 26.6 \text{ L/mol}$$

Complete Part (a).

$$D \equiv \frac{m}{V} = \frac{2.016 \text{ g}}{\text{mol}} \times \frac{1 \text{ mol}}{26.6 \text{ L}} = 0.0758 \text{ g/L}$$

Now for Part (b), finding the molar volume of dihydrogen sulfide when its density is 1.18 g/L. One of the two ratios is given. What is the other ratio you need? Find its value.

2(1.008 g/mol H) + 32.07 g/mol S = 34.09 g/mol H_2S

Write the defining equation for molar volume, and then use the two ratios to calculate its value.

$$MV \equiv \frac{L}{\text{mol}} = \frac{1 \text{ L}}{1.18 \text{ g}} \times \frac{34.09 \text{ g}}{\text{mol}} = 28.9 \text{ L/mol}$$

EXAMPLE 14.7-WB

The mass of 0.414 liter of an unknown gas at 1.07 atm and 15°C is 0.787 gram. What is the molar mass of that gas?

You need two ratios to solve this problem, density and molar volume. Density, mass per unit volume, is given in the problem statement, 0.787 gram and 0.414 liter. You could calculate the value of this ratio, but there's no need to and it's just one more opportunity to make a mistake, so let's use the ratio without doing the calculation.

The other ratio is molar volume. Determine the molar volume of a gas at 15°C and 1.07 atm.

	Molar Volume	Temperature	Pressure
Initial Value (1)	22.4 L/mol	0°C;273 K	1 atm
Final Value (2)	MV_2	15°C; 288 K	1.07 atm

$$MV_2 = MV_1 \times \frac{P_1}{P_2} \times \frac{T_2}{T_1} = 22.4 \text{ L/mol} \times \frac{1 \text{ atm}}{1.07 \text{ atm}} \times \frac{288 \text{ K}}{273 \text{ K}} = 22.1 \text{ L/mol}$$

Find the molar mass of the gas.

$$\frac{0.787 \text{ g}}{0.414 \text{ L}} \times \frac{22.1 \text{ L}}{\text{mol}} = 42.0 \text{ g/mol}$$

EXAMPLE 14.8-WB

How many grams of sodium hydroxide can be recovered as a by-product of the chemical change that produces 71.9 liters of chlorine, measured at STP, by the reaction $2\ NaCl(aq) + 2\ H_2O(\ell) \rightarrow 2\ NaOH(aq) + Cl_2(g) + H_2(g)$?

Start by listing the *GIVEN* and *WANTED* for this problem.

GIVEN: 71.9 L Cl_2 *WANTED:* g NaOH

Now write out a *PER/PATH* that takes you from liters of chlorine to grams of sodium hydroxide.

PER/PATH: L Cl_2 $\xrightarrow{22.4 \text{ L } Cl_2/\text{mol } Cl_2}$ mol Cl_2

$\xrightarrow{2 \text{ mol NaOH}/1 \text{ mol } Cl_2}$ mol NaOH $\xrightarrow{40.00 \text{ g NaOH/mol NaOH}}$ g NaOH

Complete the problem with the setup and final answer.

$$71.9 \text{ L } Cl_2 \times \frac{1 \text{ mol } Cl_2}{22.4 \text{ L } Cl_2} \times \frac{2 \text{ mol NaOH}}{1 \text{ mol } Cl_2} \times \frac{40.00 \text{ g NaOH}}{\text{mol NaOH}} = 257 \text{ g NaOH}$$

EXAMPLE 14.9-WB

How many grams of sodium hydroxide can be recovered as a by-product of the reaction that produces 71.9 liters of chlorine, measured at 1.17 atm and 38°C, by the reaction 2 NaCl(aq) + 2 $H_2O(\ell)$ → 2 NaOH(aq) + Cl_2(g) + H_2(g)?

Begin by finding the molar volume of chlorine at the given temperature and pressure.

	Molar Volume	Temperature	Pressure
Initial Value (1)	22.4 L/mol	0°C;273 K	1 atm
Final Value (2)	MV_2	38°C; 311 K	1.17 atm

$$MV_2 = MV_1 \times \frac{P_1}{P_2} \times \frac{T_2}{T_1} = 22.4 \text{ L/mol} \times \frac{1 \text{ atm}}{1.17 \text{ atm}} \times \frac{311 \text{ K}}{273 \text{ K}} = 21.8 \text{ L/mol}$$

PLAN the solution to the problem.

GIVEN: 71.9 L Cl_2 *WANTED:* g NaOH

PER/PATH: $\text{L Cl}_2 \xrightarrow{21.8\text{ L/mol}} \text{mol Cl}_2 \xrightarrow{2\text{ mol NaOH/1 mol Cl}_2} \text{mol NaOH} \xrightarrow{40.00\text{ g NaOH/mol NaOH}} \text{g NaOH}$

You are now ready to setup and solve the problem.

$$71.9\text{ L Cl}_2 \times \frac{1\text{ mol Cl}_2}{21.8\text{ L Cl}_2} \times \frac{2\text{ mol NaOH}}{1\text{ mol Cl}_2} \times \frac{40.00\text{ g NaOH}}{\text{mol NaOH}} = 264\text{ g NaOH}$$

EXAMPLE 14.10-WB

Calculate the mass of C_4H_{10} that reacts if 52.1 L O_2, measured at 0.212 atm and 24°C, is used in burning in the reaction $2\ C_4H_{10}(g) + 13\ O_2(g) \rightarrow 8\ CO_2(g) + 10\ H_2O(\ell)$.

Take this one all the way.

GIVEN: 52.1 L O_2, 0.212 atm, 24°C (297 K) *WANTED:* g C_4H_{10}

	Molar Volume	Temperature	Pressure
Initial Value (1)	22.4 L/mol	0°C;273 K	1 atm
Final Value (2)	MV_2	24°C; 297 K	0.212 atm

$$MV_2 = MV_1 \times \frac{P_1}{P_2} \times \frac{T_2}{T_1} = 22.4 \text{ L/mol} \times \frac{1 \text{ atm}}{0.212 \text{ atm}} \times \frac{297 \text{ K}}{273 \text{ K}} = 115 \text{ L/mol}$$

$$52.1 \text{ L } O_2 \times \frac{1 \text{ mol } O_2}{115 \text{ L } O_2} \times \frac{2 \text{ mol } C_4H_{10}}{13 \text{ mol } O_2} \times \frac{58.12 \text{ g } C_4H_{10}}{\text{mol } C_4H_{10}} = 4.05 \text{ g } C_4H_{10}$$

EXAMPLE 14.11-WB

How many grams of sodium hydroxide can be recovered as a by-product of the reaction that produces 71.9 liters of chlorine, measured at 1.17 atm and 38°C, by the reaction 2 NaCl(aq) + 2 $H_2O(\ell) \rightarrow$ 2 NaOH(aq) + Cl_2(g) + H_2(g)?

The first step when volume is given is to find the STP volume of that gas. Use the combined gas laws equation to do so.

GIVEN: 71.9 L Cl_2; 1.17 atm; 38°C (311 K) *WANTED:* L Cl_2 at STP

$$V_2 = V_1 \times \frac{P_1}{P_2} \times \frac{T_2}{T_1} = 71.9 \text{ L } Cl_2 \times \frac{1.17 \text{ atm}}{1 \text{ atm}} \times \frac{273 \text{ K}}{311 \text{ K}} = 73.8 \text{ L } Cl_2$$

Now you solve the problem as if it is a STP stoichiometry question.

GIVEN: 73.8 L Cl_2 at STP *WANTED:* g NaOH

PER/PATH: L Cl_2 $\xrightarrow{22.4\text{ L }Cl_2/\text{mol }Cl_2}$ mol Cl_2 $\xrightarrow{2\text{ mol NaOH}/1\text{ mol }Cl_2}$ mol NaOH $\xrightarrow{40.00\text{ g NaOH}/\text{mol NaOH}}$ g NaOH

$$73.8\text{ L }Cl_2 \times \frac{1\text{ mol }Cl_2}{22.4\text{ L }Cl_2} \times \frac{2\text{ mol NaOH}}{1\text{ mol }Cl_2} \times \frac{40.00\text{ g NaOH}}{\text{mol NaOH}} = 264\text{ g NaOH}$$

EXAMPLE 14.12-WB

Calculate the mass of C_4H_{10} that reacts if 52.1 L O_2, measured at 0.212 atm and 24°C, is used in burning in the reaction $2\ C_4H_{10}(g) + 13\ O_2(g) \rightarrow 8\ CO_2(g) + 10\ H_2O(\ell)$.

Solve the problem without further guidance.

GIVEN: 52.1 L O_2; 0.212 atm; 24°C (297 K) *WANTED:* L O_2 at STP

$$V_2 = V_1 \times \frac{P_1}{P_2} \times \frac{T_2}{T_1} = 52.1\text{ L }O_2 \times \frac{0.212\text{ atm}}{1\text{ atm}} \times \frac{273\text{ K}}{297\text{ K}} = 10.2\text{ L }O_2$$

GIVEN: 10.2 L O_2 at STP *WANTED:* mass C_4H_{10} (assume g)

PER/PATH: L O_2 $\xrightarrow{22.4\text{ L }O_2/\text{mol }O_2}$ mol O_2 $\xrightarrow{2\text{ mol }C_4H_{10}/13\text{ mol }O_2}$ mol C_4H_{10} $\xrightarrow{58.12\text{ g }C_4H_{10}/\text{mol }C_4H_{10}}$ g C_4H_{10}

$$10.2\text{ L }O_2 \times \frac{1\text{ mol }O_2}{22.4\text{ L }O_2} \times \frac{2\text{ mol }C_4H_{10}}{13\text{ mol }O_2} \times \frac{58.12\text{ g }C_4H_{10}}{\text{mol }C_4H_{10}} = 4.07\text{ g }C_4H_{10}$$

EXAMPLE 14.13-WB

A total of 3.3 L of gaseous acetylene, C_2H_2, is burned in a welding operation. If the temperature of the flame is 2955°C and the gases burn at 0.96 atm pressure, what volume of oxygen is consumed in the combustion reaction? Assume that when the oxygen reacts, it is at the same temperature and pressure as the acetylene.

Begin by writing and balancing the reaction equation.

$2\ C_2H_2 + 5\ O_2 \rightarrow 4\ CO_2 + 2\ H_2O$

Determine the volume of oxygen that reacts.

GIVEN: 3.3 L C_2H_2 *WANTED:* volume O_2 (assume L)

PER/PATH: L C_2H_2 $\xrightarrow{5\text{ L }O_2/2\text{ L }C_2H_2}$ L O_2

$$3.3\text{ L }C_2H_2 \times \frac{5\text{ L }O_2}{2\text{ L }C_2H_2} = 8.3\text{ L }O_2$$

EXAMPLE 14.14-WB

What maximum volume of NOCl(g), measured at 35°C and 0.943 atm, can be obtained from the reaction of 5.17 L Cl_2(g) at 18°C and 49.0 atm? The equation is 2 NO(g) + Cl_2(g) $\rightarrow$ 2 NOCl(g).

Start with a table of initial and final values.

	Volume	Temperature	Pressure
Initial Value (1)	5.17 L	18°C; 291 K	49.0 atm
Final Value (2)	V_2	35°C; 308 K	0.943 atm

Determine the volume occupied by chlorine gas at the final conditions.

$$V_2 = V_1 \times \frac{P_1}{P_2} \times \frac{T_2}{T_1} = 5.17 \text{ L} \times \frac{49.0 \text{ atm}}{0.943 \text{ atm}} \times \frac{308 \text{ K}}{291 \text{ K}} = 284 \text{ L}$$

You can now complete the problem by setting up and solving the one-step conversion to volume of NOCl(g).

GIVEN: 284 L Cl_2 *WANTED:* volume NOCl (assume L)

PER/PATH: L Cl_2 $\xrightarrow{2 \text{ L NOCl}/1 \text{ L Cl}_2}$ L NOCl

$$284 \text{ L Cl}_2 \times \frac{2 \text{ L NOCl}}{1 \text{ L Cl}_2} = 568 \text{ L NOCl}$$

Name: ______________________ Date: ______________________

ID: ______________________ Section: ______________________

Questions, Exercises, and Problems

Section 14.2: Molar Volume

1. Explain the restrictions placed on the statement that 22.4 L/mol is the molar volume of any gas.

2. What volume will be occupied by 4.21 moles of carbon monoxide at STP?

Answer: [ ]

3. Compare the molar volumes of helium and neon at 30°C and 1.10 torr. What are their values in L/mol?

Answer: [ ]

4. Find the molar volume of acetylene, C_2H_2, at 21°C and 0.908 atm.

Answer: [ ]

Section 14.3: Three Ratios: Gas Density, Molar Volume, and Molar Mass

5. What volume is occupied by 9.08 grams of carbon monoxide at a temperature and pressure at which its density is 0.605 g/L?

Answer: [ ]

Name: ______________________ Date: ______________________

ID: ______________________ Section: ______________________

6. On a cold winter day the temperature is 0°F (–18°C) and the density of air is 1.39 g/L. Calculate the mass of air in a 1-gallon (3.79-L) bottle at those conditions.

Answer: ____________

Section 14.4: Three-Ratio Problems

7. Calculate the density of helium (a) at STP and (b) at 0.76 atm and –104°C.

Answer: ____________

8. The density of neon is 4.63 g/L in a pressurized tank. Calculate the molar volume of the gas in the tank. If the volume of the tank is 2.43 liters, how many moles of neon are in the tank?

Answer: ____________

9. If the density of an unknown gas at 41°C and 2.61 atm is 1.61 g/L, what is its molar mass?

Answer: ____________

Name: ______________________ Date: ______________________

ID: ______________________ Section: ______________________

10. Nitrogen accounts for nearly 80% of the gaseous mixture that makes up the atmosphere. What is the density of pure nitrogen at 0.800 atm and 15°C?

Answer: [ ]

11. The gas in an 8.07-liter cylinder at 13°C has a mass of 33.5 grams and exerts a pressure of 3.25 atm. Find the molar mass of the gas.

Answer: [ ]

12. Calculate the density of ethylene, C_2H_4, at 914 torr and 37°C.

Answer: [ ]

13. NO_2 and N_2O_4 both have the same empirical (simplest) formula. At a temperature and pressure at which both substances are gases, can you tell without calculating which gas is more dense? Explain.

Name: ______________________ Date: ______________________

ID: ______________________ Section: ______________________

Section 14.5: Gas Stoichiometry at Standard Temperature and Pressure (STP)

14. One small-scale laboratory method for preparing oxygen is to heat potassium chlorate in the presence of a catalyst: $2\ KClO_3(s) \rightarrow 2\ KCl(s) + 3\ O_2(g)$. Find the STP volume of oxygen that can be produced by 5.74 g $KClO_3$.

Answer: [ ]

15. The reaction used to produce chlorine in the laboratory is $2\ KMnO_4(aq) + 16\ HCl(aq) \rightarrow 2\ MnCl_2(aq) + 5\ Cl_2(aq) + 8\ H_2O(aq) + 2\ KCl(aq)$. Calculate the number of grams of potassium permanganate, $KMnO_4$, that are needed to produce 9.81 L of chlorine, measured at STP.

Answer: [ ]

Section 14.6: Gas Stoichiometry: Molar Volume Method
Section 14.7: Gas Stoichiometry: Combined Gas Equation Method

Questions 16–19 may be solved by the molar volume method (Section 14.6) or by the combined gas equation method (Section 14.7). Setup your work according to the section you studied.

16. One source of sulfur dioxide used in making sulfuric acid comes from sulfide ores by the reaction $4\ FeS_2(s) + 11\ O_2(g) \rightarrow 2\ Fe_2O_3(s) + 8\ SO_2(g)$. How many liters of SO_2, measured at 983 torr and 214°C, are produced by the reaction of 598 g FeS_2?

Answer: [ ]

Name: ______________________ Date: ______________________

ID: ______________________ Section: ______________________

17. How many grams of water must decompose by electrolysis to produce 23.9 L H_2, measured at 28°C and 728 torr?

Answer: ______________________

18. The reaction chamber in a modified Haber process for making ammonia by the direct combination of its elements is operated at 575°C and 248 atm. How many liters of nitrogen, measured at these conditions, will react to produce 9.16×10^3 grams of ammonia?

Answer: ______________________

19. When properly detonated, ammonium nitrate explodes violently, releasing hot gases: $NH_4NO_3(s) \rightarrow N_2O(g) + 2\ H_2O(g)$. If the total volume of gas produced, both dinitrogen and steam, is 82.3 L at 447°C and 896 torr, how many grams of NH_4NO_3 exploded?

Answer: ______________________

Name: ______________________ Date: ______________________

ID: ______________________ Section: ______________________

Section 14.8: Volume–Volume Gas Stoichiometry

20. Compare the volumes of 1×10^{23} hydrogen molecules, 1×10^{23} oxygen molecules, and 2×10^{23} nitrogen molecules, all at the same temperature and pressure.

21. Sulfur burns to SO_2 with a beautiful deep blue–purple flame, but with a foul, suffocating odor: $2\ S + O_2 \rightarrow 2\ SO_2$. (a) How many liters of O_2 are needed to form 35.2 L SO_2, both gases being measured at 741 torr and 26°C? (b) What if only the SO_2 is at those conditions, but the O_2 is at 17°C and 847 torr?

Answer: ____________

22. Gaseous chlorine dioxide, ClO_2, is used to bleach flour and in water treatment. It is produced by the reaction of chlorine with sodium chlorite: $Cl_2 + 2\ NaClO_2 \rightarrow 2\ ClO_2 + 2\ NaCl$. How many liters of ClO_2, measured at 0.961 atm and 31°C, will be produced by 283 L Cl_2 at 2.91 atm and 21°C?

Answer: ____________

Name: ______________________ Date: ______________________

ID: ______________________ Section: ______________________

23. In the natural oxidation of hydrogen sulfide released by decaying organic matter, the following reaction occurs: $2\ H_2S + 3\ O_2 \rightarrow 2\ SO_2 + 2\ H_2O$. How many liters of hydrogen sulfide, measured at 19°C and 549 torr, will be used in a reaction that also uses 704 mL O_2 at 159 torr and 26°C?

Answer: ____________

General Questions

24. Find the mass of 57.9 liters of krypton at 775 torr and 6°C.

Answer: ____________

25. At 17°C and 0.835 atm, 16.2 liters of ammonia has a mass of 9.68 g. What is the molar volume of ammonia at those conditions? (This is easier than it may seem!)

Answer: ____________

Name: ____________________ Date: ____________________

ID: ____________________ Section: ____________________

26. A 7.60-g sample of pure liquid is vaporized at 183°C and 179 torr. At these conditions it occupies 3.87 L. What is the molar mass of the substance?

Answer: [ ]

More Challenging Problems

27. The density of nitrogen at 0.913 atm and 18°C is 1.07 g/L. Explain how this shows that the formula of nitrogen is N_2 rather than just N.

28. An organic chemist has produced a solid she believes to be pure; she expects a molar mass of 346 g/mol. Using 4.08 grams of the solid, she melts and then boils it in a 3.36-L vacuum chamber at 117 torr and 243°C. She is disappointed; the molar mass is close to what she expected, but not close enough. (a) What molar mass did she find? (b) Her finding suggested to her what her mistake might have been. Does it suggest anything to you? [Part (b) is beyond the scope of Chapter 13, but you might have an inspiration if you have studied Chapter 12.)

Name: ______________________ Date: ______________________

ID: ______________________ Section: ______________________

29. Air is a mixture. Consequently, its physical properties vary from day to day, depending on the levels of moisture, particulates, and other substances in the atmosphere. On a particular day, the density of air is found to be 1.3 g/L at STP. Use that density to calculate the "effective" molar mass of that sample of air, that is, the molar mass that a pure gas would have if its STP density were 1.3 g/L. Then use the calculated molar mass to find what the density would be at 20°C and 751 torr.

Answer: [ ]

30. A student evacuates a gas-weighing bottle and finds its mass to be 135.831 g. She then fills the bottle with an unknown gas, adjusts the temperature and pressure to STP, and weighs it again. Its mass is now 136.201 g. She then fills the bottle with water and finds its mass to be 385.42 g. Calculate the molar mass of the gas. (Recall the density of water is 1.00 g/mL.)

Answer: [ ]

Answers to Target Checks

1. $$MV \equiv \frac{V}{n} = \frac{RT}{P} = \frac{62.4\ L \cdot torr}{mol \cdot K} \times 326\ K \times \frac{1}{693\ torr} = 29.4\ L/mol$$

$$10.0\ L\ N_2 \times \frac{1\ mol\ N_2}{29.4\ L\ N_2} = 0.340\ mol\ N_2$$

Chapter 15

Gases, Liquids, and Solids

EXAMPLE 15.1-WB

The total pressure in an oxygen generator such as that shown in Figure 15.2 in the textbook is 755 torr. The temperature of the system is 22°C, at which the water vapor pressure is 19.8 torr. What is the partial pressure of the oxygen?

Complete the problem.

$P = p_{oxygen} + p_{water}$

$p_{oxygen} = P - p_{water} = 755 \text{ torr} - 19.8 \text{ torr} = 735 \text{ torr}$

If you did not get the correct number of significant figures, you may want to review the "Addition and Subtraction" discussion in Section 3.5.

☑ TARGET CHECK 15.1-WB

a) Explain why intermolecular forces are weaker in gases than in liquids.

b) Equal volumes of liquids A and B, at the same temperature, are poured through identical funnels. Each funnel has a long, narrow stem. Liquid A requires more time to flow through the funnel than liquid B. Compare the intermolecular attractions in A and B. What liquid property is being described? Write a definition of that property.

☑ TARGET CHECK 15.2-WB

The term *hydrogen bond* makes it sound as if this intermolecular force is similar to a covalent bond, ionic bond, or metallic bond. Explain why a hydrogen bond is not actually a chemical bond.

☑ TARGET CHECK 15.3-WB

Describe the intermolecular forces in a sample of carbon tetrachloride that contribute to the fact that it exists as a liquid at room conditions.

☑ TARGET CHECK 15.4-WB

Compare the boiling points of SiH_3Cl, CH_3Cl, and CH_4, as would be predicted based on the relative strengths of the intermolecular forces present in samples of each substance. Explain your reasoning.

☑ TARGET CHECK 15.5-WB

A highly volatile liquid, CH_2Cl_2, is placed in a container. The container is then sealed. Some, but not all of the liquid disappears. How does the vapor pressure of the CH_2Cl_2 in the container compare with the equilibrium vapor pressure? Explain your reasoning.

☑ TARGET CHECK 15.6-WB

Why does a properly stated boiling point include both a boiling temperature and the surrounding pressure? What pressure is implied in the everyday language usage of the term "boiling point"?

☑ TARGET CHECK 15.7-WB

Both window glass and quartz are made up largely of silicon dioxide. Window glass softens over a wide temperature range, while quartz melts at 1610°C. Which solid, window glass or quartz, is most likely an amorphous solid? Which is most likely a crystalline solid? Explain your reasoning.

☑ TARGET CHECK 15.8-WB

Substance A melts at 2810°C and is insoluble in water. It conducts electricity when melted, but not when solid. Analyze each fact given, and explain how it is supporting or refuting evidence that Substance A is an ionic, molecular, network, or metallic crystal. Which crystal type is most likely for Substance A, based on these data?

EXAMPLE 15.2-WB

A researcher determines that 744 J of energy is needed to vaporize a 355-mg sample of a liquid at its boiling point. Determine the heat of vaporization and the heat of condensation of the substance. Express both values in kJ/g.

What is the relationship between heat of vaporization and heat of condensation for a pure substance?

They have the same value. Heat of condensation values are negative; heat of vaporization values are positive.

Once you have determined the value of the heat of vaporization, you have also found the heat of condensation value—all you need to do is add a negative sign. Write the definition of heat of vaporization, substitute the values, make the necessary conversions, and solve. Also write the value of the heat of condensation.

GIVEN: 744 J; 355 mg *WANTED:* ΔH_{vap} in kJ/g

$$\Delta H_{vap} \equiv \frac{q}{m} = \frac{744\ J}{355\ mg} \times \frac{1\ kJ}{1000\ J} \times \frac{1000\ mg}{g} = 2.10\ kJ/g$$

$$\Delta H_{condensation} = -2.10\ kJ/g$$

EXAMPLE 15.3-WB

Calculate the energy required to vaporize 255 grams of water at its boiling point.

Refer to Table 15.4 on Page 418, and write the *GIVEN* quantities and *WANTED* units.

GIVEN: 255 g; $\Delta H_{vap} = 2.26$ kJ/g *WANTED:* Energy (assume kJ)

In most "real world" problems, you will have to use a reference source to find values of data such as heats of vaporization.

Set up and solve.

$\Delta H_{vap} \equiv \frac{q}{m}$, so cross multiply to get $q = \Delta H_{vap} \times m$

$$q = \Delta H_{vap} \times m = \frac{2.26 \text{ kJ}}{\text{g}} \times 255 \text{ g} = 576 \text{ kJ}$$

EXAMPLE 15.4-WB

How many grams of lead can be melted by 749 J of heat energy?

Find the needed "missing information," set up, and solve.

GIVEN: 749 J; 23 J/g *WANTED:* g

$\Delta H_{fus} \equiv \frac{q}{m}$, so cross multiply to get $q = \Delta H_{fus} \times m$, and then divide both sides by ΔH_{fus}:

$$m = \frac{q}{\Delta H_{fus}} = q \times \frac{1}{\Delta H_{fus}} = 749 \text{ J} \times \frac{1 \text{ g}}{23 \text{ J}} = 33 \text{ g}$$

EXAMPLE 15.5-WB

A student observes a loss of 803 joules as 141 grams of aluminum cools from 31.7°C to 25.0°C. Calculate the specific heat of aluminum from these data.

PLAN the problem. Be careful about the value of q.

GIVEN: –803 J; 141 g; T_i = 31.7°C; T_f = 25.0°C *WANTED:* c in J/g · °C

EQUATION: $q = m \times c \times \Delta T$

Notice that q is a negative quantity. This is because aluminum is losing energy, as seen by the drop in temperature.

You are now ready to solve the equation for the wanted variable, substitute the given quantities, and calculate the answer. You may calculate ΔT and substitute the result into the solved equation, or you may replace ΔT by its source, $T_f - T_i$. Be careful about algebraic signs.

$$c = \frac{q}{m \times \Delta T} = \frac{q}{m \times (T_f - T_i)} = -803\text{ J} \times \frac{1}{141\text{ g}} \times \frac{1}{(25.0 - 31.7)°\text{C}} = 0.85\text{ J/g} \cdot °\text{C}$$

ΔT is 25.0°C – 31.7°C = –6.7°C. This negative value combines with the negative value of q to yield a positive specific heat value. Note that even though every number in the data is a three-significant-figure number, the *change* in temperature has only two significant figures. This limits the answer to two significant figures.

EXAMPLE 15.7-WB

What is the total heat flow if 114 grams of ice, $H_2O(s)$, initially at –81°C warms to water at 39°C? The specific heat of ice is 2.1 J/g · °C; the specific heat of water is 4.18 J/g · °C. The heat of fusion of water is 335 J/g, and the melting point of water is 0°C.

Sketch a temperature–heat curve for this problem.

Your curve should have three sections:
(1) a solid section with a positive slope that ranges from –81°C to 0°C
(2) a solid + liquid section with zero slope at 0°C
(3) a liquid section with a positive slope that ranges from 0°C to 39°C
See the first three sections of the curve in Figure 15.24 on Page 423 for a similar example.

You have three separate q calculations to carry out. Start with the *PLAN* for the calculation for the heat flow into the solid.

GIVEN: $m = 114\text{ g}$; $c = 2.1\text{ J/g}\cdot{}^\circ\text{C}$; $T_f = 0^\circ\text{C}$; $T_i = -81^\circ\text{C}$ *WANTED:* q (assume kJ)

EQUATION: $q = m \times c \times \Delta T$

Carry out the calculation. Note that we made the assumption that the heat flow should be expressed in kilojoules. The kilounit is often used in change in temperature plus change of state problems because the total heat flow is usually more than 1000 joules.

$$q = m \times c \times \Delta T = 114\text{ g} \times \frac{2.1\text{ J}}{\text{g}\cdot{}^\circ\text{C}} \times [0-(-81)]^\circ\text{C} \times \frac{1\text{ kJ}}{1000\text{ J}} = 19\text{ kJ}$$

Both the specific heat value and the ΔT limit the answer to two significant figures.

Now *PLAN* and calculate the heat flow needed to melt the solid.

GIVEN: $m = 114\text{ g}$; $\Delta H_{fus} = 335\text{ J/g}$ *WANTED:* q in kJ

EQUATION: $q = m \times \Delta H_{fus} = 114\text{ g} \times \frac{335\text{ J}}{\text{g}} \times \frac{1\text{ kJ}}{1000\text{ J}} = 38.2\text{ kJ}$

The third and final heat flow calculation involves the heat needed to warm the water. Go all the way to the answer for this part of the curve.

GIVEN: m = 114 g; c = 4.18 J/g · °C; T_f = 39°C; T_i = 0°C *WANTED:* q in kJ

$$q = m \times c \times \Delta T = 114\ g \times \frac{4.18\ J}{g \cdot °C} \times (39 - 0)°C \times \frac{1\ kJ}{1000\ J} = 19\ kJ$$

Complete the problem by finding the total heat flow. Remember to apply the addition/subtraction rule for significant figures.

Σq = 19 kJ + 38.2 kJ + 19 kJ = 76 kJ

The final answer is expressed to the ones place because two of the added quantities are known to that accuracy.

Name: ______________________ Date: ______________________

ID: ______________________ Section: ______________________

Questions, Exercises, and Problems

Section 15.1: Dalton's Law of Partial Pressures

1. A mixture of helium and argon occupies 1×10^2 L. The partial pressure of helium is 0.6 atm, and the partial pressure of argon is 0.4 atm. What are the partial volumes of the two gases? Explain your answer.

2. Atmospheric pressure is the total pressure of the gaseous mixture called air. Atmospheric pressure is 749 torr on a day that the partial pressures of nitrogen, oxygen, and argon are 584 torr, 144 torr, and 19 torr, respectively. What is the partial pressure of all the other gases in the air on that day?

Answer: ______________________

3. A sample of "wet" hydrogen gas was collected over water (Figure 15.2). The sample was adjusted so that it was at room temperature and pressure, 19°C and 733 mm Hg. Water vapor pressure at 19°C is 16.5 mm Hg. What is the partial pressure of the "dry" hydrogen gas?

Answer: ______________________

Section 15.2: Properties of Liquids

4. Why will two gases mix with each other more quickly than two liquids?

5. Why are intermolecular attractions stronger in the liquid state than in the gaseous state?

Name: ______________________ Date: ______________________

ID: ______________________ Section: ______________________

6. How do intermolecular attractive forces influence the boiling point of a pure substance?

7. Why does molar heat of vaporization depend on the strength of intermolecular forces?

8. A tall glass cylinder is filled to a depth of 1 meter with water. Another tall glass cylinder is filled to the same depth with syrup. Identical ball bearings are dropped into each tube at the same instant. In which tube will the ball bearings reach the bottom first? Explain your prediction in terms of viscosity and intermolecular attractive forces.

9. If water is spilled on a laboratory desktop, it usually spreads over the surface, wetting any papers or books that may be in its path. If mercury is spilled, it neither spreads nor makes paper wet, but forms little drops that are easily combined into pools by pushing them together. Suggest an explanation for these facts in terms of the apparent surface tension and intermolecular attractive forces in mercury and water.

10. The level at which a duck floats on water is determined more by the thin oil film that covers its feathers than by a body density that is lower than the density of water. The water does not "mix" with the oil, and therefore does not penetrate the feathers. If, however, a few drops of "wetting agent" are placed in the water near the duck, the poor duck will sink to its neck. State the effect of a wetting agent on surface tension and intermolecular attractions of water.

Name: ______________________ Date: ______________________

ID: ______________________ Section: ______________________

Questions 11 and 12: The table below gives the normal boiling and melting points for three nitrogen oxides.

	NO	**N_2O**	**NO_2**
Boiling point	–152°C	–88.5°C	+21.2°C
Melting point	–164°C	–90.8°C	–11.2°C

11. Which of the three oxides would you expect to have the highest molar heat of vaporization? Explain how you reached your conclusion.

12. Which of the three oxides would you expect to have a measurable vapor pressure at –90°C? Explain your answer.

Section 15.3: Types of Intermolecular Forces

13. Other things being equal, which produces stronger intermolecular attractions, induced dipole forces or dipole forces? What "other things being *un*equal" would reverse this order of attractions?

14. What are the principal intermolecular forces in each of the following compounds: $NH(CH_3)_2$, CH_2F_2, C_3H_8?

Name: ____________________ Date: ____________________

ID: ____________________ Section: ____________________

15. Compare dipole forces and hydrogen bonds. How are they different, and how are they similar?

Questions 16 and 17: On the basis of molecular size, molecular polarity, and hydrogen bonding, predict for each pair of compounds the one that has the higher boiling point. State the reason for your choice. Assume molecular size is related to molar mass.

16. CH_4 and NH_3

17. Ar and Ne

18. What feature of the hydrogen atom, when bonded to an appropriate second element, is largely responsible for the strength of hydrogen bonding between molecules?

19. Of the three types of intermolecular forces, which one(s) (a) increase with molecular size; (b) account for the high melting point, boiling point, and other abnormal properties of water?

Name: ______________________ Date: ______________________

ID: ______________________ Section: ______________________

20. Identify the intermolecular forces present in each of the following:

a)

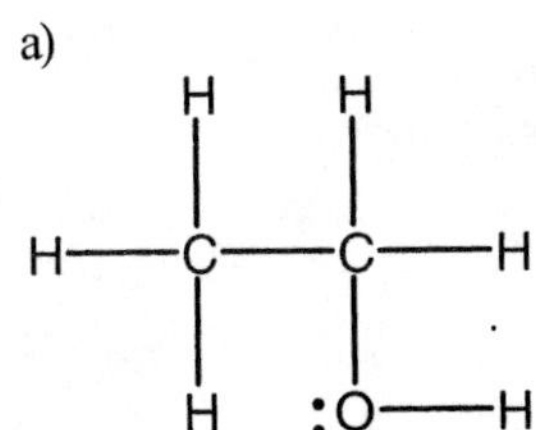

b)

c)

d)

e)

Name: ____________________ Date: ____________________

ID: ____________________ Section: ____________________

21. Predict which compound, CO_2 or CS_2, has the higher melting and boiling points. Explain your prediction.

22. Predict which compound, CH_4 or CH_3F, has the higher vapor pressure as a liquid at a given temperature. Explain your prediction.

Section 15.4: Liquid–Vapor Equilibrium

23. What is the meaning of *equilibrium*?

24. Explain why the rate of evaporation from a liquid depends on temperature.

Section 15.5: The Boiling Process

25. The vapor pressure of a certain compound at 20°C is 906 torr. Is the substance a gas or a liquid at 760 torr? Explain.

Name: ______________________ Date: ______________________

ID: ______________________ Section: ______________________

26. Liquid feed water is delivered to modern boilers at a temperature well above the normal boiling point of water. Explain how this is possible.

27. Explain why low boiling liquids usually have low molar heats of vaporization.

28. At 20°C the vapor pressure of substance M is 520 torr; of substance N, 634 torr. Which substance will have the lower boiling point? the lower molar heat of vaporization?

Section 15.7: The Solid State

29. Is ice a crystalline solid or an amorphous solid? On what properties do you base your conclusion?

Section 15.8: Types of Crystalline Solids

30. The physical properties of two solids are tabulated below. In each case, state whether the solid is most likely to be ionic, molecular, metallic, or a covalent network solid.

Solid	Melting Point	Water Solubility	Conductivity (Pure)	Type of Solid
C	2000°C	Insoluble	Nonconductor	______
D	1050°C	Soluble	Nonconductor	______

Name: ______________________________ Date: ______________________

ID: ________________________________ Section: ____________________

Section 15.9: Energy and Change of State

See Table 15.4 on Page 418 of the textbook for heats of fusion and vaporization.

31. A student is to find the heat of vaporization of isopropyl alcohol (rubbing alcohol). She vaporizes 61.2 g of the liquid at its boiling point and measures the energy required as 44.8 kJ. What heat of vaporization does she report?

Answer: [ ]

32. Calculate the energy released as 227 grams of sodium vapor condenses.

Answer: [ ]

33. 79.4 kJ was released by the condensation of a sample of ethyl alcohol. If $\Delta H_{vap} = 0.880$ kJ/g, what was the mass of the sample?

Answer: [ ]

34. Acetone, C_3H_6O, is a highly volatile solvent sometimes used as a cleansing agent prior to vaccination. It evaporates quickly from the skin, making the skin feel cold. How much energy is absorbed by 23.8 g of acetone as it evaporates if its molar heat of vaporization is 32.0 kJ/mol?

Answer: [ ]

Name: ____________________ Date: ____________________

ID: ____________________ Section: ____________________

35. Calculate the energy lost when 3.30 kg of lead freeze.

Answer: [ ]

36. 36.9 g of an unknown metal releases 2.51 kJ of energy in freezing. What is the heat of fusion of that metal?

Answer: [ ]

37. A piece of zinc releases 4.45 kJ while freezing. What is the mass of the sample?

Answer: [ ]

Section 15.10: Energy and Change of Temperature: Specific Heat

See Table 15.5 on Page 421 of the textbook for specific heat values.

38. Samples of two different metals, A and B, have the same mass. Both samples absorb the same amount of energy. The temperature of A increases by 11°C, and the sample of B increases by 13°C. Which metal has the higher specific heat? Explain your reasoning.

Name: ______________________ Date: ______________________

ID: ______________________ Section: ______________________

39. Find the quantity of energy released (in joules) as 467 grams of zinc cool from 68°C to 31°C.

Answer: [ ]

40. How much energy (kJ) is required to cool 2.30 kilograms of gold from 88°C to 22°C?

Answer: [ ]

41. The mass of a handful of copper coins is 144 grams. The coins are at a temperature of 33°C. If they lose 1.47 kJ when they are tossed in a fountain and drop to the fountain's water temperature, what is that temperature?

Answer: [ ]

Section 15.11: Change in Temperature Plus Change of State

Questions 42 to 46: Figure 15.25 on textbook Page 428 is a graph of temperature versus energy for a sample of a pure substance. Assume that letters J through P on the horizontal and vertical axes represent numbers and that expressions such as R – S or X + Y + Z represent arithmetic operations to be performed with those numbers.

42. Identify by letter the boiling and freezing points in Figure 15.25.

Name: ______________________ Date: ______________________

ID: ______________________ Section: ______________________

43. Identify all points on the curve in Figure 15.25 where the substance is entirely gas.

44. Identify in Figure 15.25 all points on the curve where the substance is partly solid and partly liquid.

45. Describe the physical changes that occur as energy N – P is removed from the sample.

46. Using letters from the graph, show how you would calculate the energy required to boil the liquid at its boiling point.

47. A 127 gram piece of ice is removed from a refrigerator at –11°C. It is placed in a bowl where it melts and eventually warms to room temperature, 21°C. Calculate the amount of energy the sample has gained from the atmosphere.

Answer: ______________________

Name: ______________________ Date: ______________________

ID: ______________________ Section: ______________________

48. A home melting pot is used for a metal casting hobby. At the end of a work period, the pot contains 689 grams of zinc at 552°C. How much energy will be released as the molten metal cools, solidifies, and cools further to room temperature, 21°C? Find the necessary data from the tables in the textbook.

Answer: ______________________

49. A certain "white metal" alloy of lead, antimony, and bismuth melts at 264°C, and its heat of fusion is 29 J/g. Its average specific heat is 0.21 J/g · °C as a liquid and 0.27 J/g · °C as a solid. How much energy is required to heat 941 kg of the alloy in a melting pot from a starting temperature of 26°C to its operating temperature, 339°C?

Answer: ______________________

Name: ______________________ Date: ______________________

ID: ______________________ Section: ______________________

General Questions

50. Under what circumstances might you find that a substance having only induced dipole forces is more viscous that a substance that exhibits hydrogen bonding?

More Challenging Problems

51. Three closed containers have identical volumes. A beaker containing a large quantity of ether, a highly volatile liquid, is placed in Container A. It evaporates until equilibrium is reached with a substantial amount of ether remaining. A beaker with a small amount of ether is placed in Container B. The ether all evaporates. A beaker with an intermediate amount of ether is placed in Container C. It evaporates until it reaches equilibrium with only a small amount of ether remaining. Compare the final either vapor pressures in the three containers. Explain your answer.

52. The equilibrium vapor pressure of water is 24°C is 22.4 torr. A sealed flask contains air at 24°C and 757 torr and a glass vial filled with liquid water. The vial is broken, allowing some of the water to evaporate. What is the maximum pressure this system can reach?

53. An industrial process requires boiling a liquid whose boiling point is so high that maintenance costs on associated pumping equipment are prohibitive. Suggest a way this problem might be solved.

Name: ______________________ Date: ______________________

ID: ______________________ Section: ______________________

54. A 54.1 g aluminum ice tray in a home refrigerator holds 408 g of water. Calculate the energy that must be removed from the tray and its contents to reduce the temperature from 17°C to 0°C, freeze the water, and drop the temperature of the tray and ice to –9°C. Assume the specific heat of aluminum remains constant over the temperature range involved.

Answer: ______________________

55. The melting point of an amorphous solid is not always a definite value as it should be for a pure substance. Suggest a reason for this.

56. It is a hot summer day and Chris wants a glass of lemonade. There is none in the refrigerator, so a new batch is prepared from freshly squeezed lemons. When finished, there are 175 grams of lemonade at 23°C. That is not a very refreshing temperature, so it must be cooled with ice. But Chris doesn't like ice in lemonade! Therefore, just enough ice is used to cool the lemonade to 5°C. Of course, the ice will melt and reach the same temperature. If the ice starts at –8°C, and if the specific heat of lemonade is the same as that of water, how many grams of ice does Chris use? Assume there is no heat transfer to or from the surroundings. Answer in two significant figures.

Answer: ______________________

Answers to Target Checks

1. a) Gas particles are so far apart that the attractive forces between them are negligible. In a liquid, the particles are close to one another, so the attractions are strong enough to have an effect on its properties.

   b) The intermolecular attractions are stronger in A than in B. The ability of a liquid to flow is measured by its viscosity, an internal resistance to flow.

2. The term *chemical bond* refers to the forces that hold atoms together in metals (metallic bonds), or to form molecules or ions (covalent bonds), or that hold oppositely charged ions together (ionic bonds). These are relatively strong forces. Intermolecular forces, such as hydrogen bonds, are much weaker forces that act between molecules. A hydrogen bond is an intermolecular force, not a chemical bond.

3. Although the C–Cl bonds in CCl_4 are polar, they are symmetrically arranged in its tetrahedral molecular geometry, so the molecule itself is net nonpolar. The only intermolecular force acting between the molecules is therefore induced dipoles.

4. $SiH_3Cl > CH_3Cl > CH_4$. SiH_3Cl and CH_3Cl have dipole forces as the primary intermolecular force, and CH_4 has induced dipoles as the primary intermolecular force. SiH_3Cl is heavier than CH_3Cl and therefore it can be assumed to be larger, resulting in a higher boiling point.

5. The vapor pressure of the CH_2Cl_2 in the container *is* the equilibrium vapor pressure. The process described is similar to that illustrated in Figure 15.11. The equilibrium vapor pressure is the partial pressure exerted by a vapor in equilibrium with its liquid phase.

6. The boiling point is that temperature at which the vapor pressure of the liquid is equal to the pressure above its surface. Both the temperature and external pressure are needed for a complete description of boiling point. The everyday usage of the term "boiling point" refers to the *normal boiling point,* which is the temperature at which a substance boils when the external pressure is one atmosphere.

7. The melting point of window glass is not sharply defined, which implies that it does not have an orderly arrangement at the particulate level, and thus different regions melt at different temperatures. It is an amorphous solid. The sharp melting point of quartz indicates that there is a particulate-level geometric pattern that is consistent throughout the solid. It is a crystalline solid.

8. • Melts at 2810°C: This is a relatively high melting point, which is consistent with ionic and network solids, and possibly with metallic solids, but not molecular solids.
   • Insoluble in water: Consistent with metallic and network solids. Probably not an ionic or molecular solid, although these are not absolutely ruled out.
   • Conducts electricity when melted: This is characteristic of ionic solids, but not the others.
   • Does not conduct when solid: Not metallic. Could be ionic, molecular, or network.
   Conclusion: Substance A is an ionic solid. All other types were excluded by the conductivity of the liquid.

Chapter 16

Solutions

☑ TARGET CHECK 16.1-WB

Are the contents of a freshly opened bottle of root beer a solution? Write the definition of the term *solution,* and explain how root beer compares with each part of the definition. Does it make any difference if the root beer is flat?

☑ TARGET CHECK 16.2-WB

a) Colored sugar water, as found in hummingbird feeders, is an example of a solution. It is mainly water with dissolved solid sugar and a few drops of liquid red food coloring. Name the solvent and solute(s) in this solution.

b) How can you make a dilute solution from a concentrated solution? How can you make a concentrated solution from a dilute solution?

c) You can dissolve 14 g of solute Y in 100 g water to form a saturated solution. Consider a scale ranging from 0 g Y in 100 g water to 30 g Y in 100 g water.
i) What range of masses of Y will constitute a saturated solution?
ii) What range of masses of Y will constitute a unsaturated solution?
iii) What range of masses of Y will constitute a supersaturated solution?

d) Gasoline and water do not dissolve in one another. Are they miscible or immiscible liquids?

☑ TARGET CHECK 16.3-WB

a) Making candy often calls for dissolving a large amount of sugar in water to form a saturated sugar solution. Both the sugar crystal and the liquid water have dipole forces and hydrogen bonding as intermolecular attractions. Describe this dissolving process in words.

b) Explain how you could speed up the dissolving process described in part (a).

☑ TARGET CHECK 16.4-WB

a) Diesel cars are usually equipped with water traps, to remove water from diesel fuel. Diesel fuel is a hydrocarbon mixture with an average formula of $C_{20}H_{42}$. Predict if water and diesel fuel will mix, and state your reasons for your prediction.

b) The partial pressure of CO_2 in the atmosphere is about 0.3 torr. If this partial pressure increases to 0.6 torr, how would bottlers change the amount of CO_2 in their carbonated beverages to maintain constant "fizziness"?

EXAMPLE 16.1-WB

3.50 g KNO_3 is dissolved in 25.0 g H_2O. Calculate the percentage concentration of KNO_3.

GIVEN: 3.50 g KNO_3; 25.0 g H_2O *WANTED*: % KNO_3

EQUATION: $\% \text{ by mass} = \dfrac{\text{g solute}}{\text{g solute} + \text{g solvent}} \times 100 = \dfrac{3.50 \text{ g solute}}{(3.50 \text{ g} + 25.0 \text{ g}) \text{ solution}} \times 100 = 12.3\% \ KNO_3$

EXAMPLE 16.2-WB

What mass of water and calcium chloride would you need to make 1562 g of a 5.03% calcium chloride solution?

The percentage concentration by mass given, 5.03% calcium chloride, can be written as a *PER* expression. Write it so that you can more clearly see the solution *PATH.*

5.03 g $CaCl_2$/100 g solution

When working with solution concentration problems, it is important to think about the given concentrations as a ratio. You will often use the ratio to convert between the units found in the ratio.

Now write the *PLAN* for finding the mass of calcium chloride in the solution.

GIVEN: 1562 g solution *WANTED*: mass $CaCl_2$ (assume grams)

PER/PATH: g solution $\xrightarrow{5.03 \text{ g } CaCl_2/100 \text{ g solution}}$ g $CaCl_2$

Calculate the mass of calcium chloride in the solution.

$$1562 \text{ g solution} \times \frac{5.03 \text{ g CaCl}_2}{100 \text{ g solution}} = 78.6 \text{ g CaCl}_2$$

The total mass of the solution is 1562 g. The mass of calcium chloride is 78.6 g. What is the mass of water?

$$\text{g H}_2\text{O} = \text{g solution} - \text{g solute} = 1562 \text{ g} - 78.6 \text{ g} = 1483 \text{ g H}_2\text{O}$$

EXAMPLE 16.3-WB

What is the molarity of the solution made by dissolving 12.9 g solid sodium bromide in water and diluting to a final volume of 5.00×10^2 mL?

The definition of molarity is a key to solving this problem. Write it.

$$M \equiv \frac{\text{mol}}{\text{L}}$$

Your task is to find the numerator and denominator in the molarity ratio. First, you need the moles of solute. *PLAN* and solve this part of the problem.

GIVEN: 12.9 g NaBr *WANTED*: mol NaBr

PER/PATH: g NaBr $\xrightarrow{102.89 \text{ g NaBr/mol NaBr}}$ mol NaBr

$$12.9 \text{ g NaBr} \times \frac{1 \text{ mol NaBr}}{102.89 \text{ g NaBr}} = 0.125 \text{ mol NaBr}$$

Now you need the value of the denominator, volume of solution in liters.

GIVEN: 5.00×10^2 mL *WANTED:* L

PER/PATH: mL $\xrightarrow{1000\ \text{mL/L}}$ L

$$5.00 \times 10^2\ \text{mL} \times \frac{1\ \text{L}}{1000\ \text{mL}} = 0.500\ \text{L}$$

Now you have both parts of the molarity ratio. Complete the final calculation.

$$\text{M} \equiv \frac{\text{mol}}{\text{L}} = \frac{0.125\ \text{mol NaBr}}{0.500\ \text{L}} = 0.250\ \text{mol NaBr/L} = 0.250\ \text{M NaBr}$$

EXAMPLE 16.4-WB

How many grams of potassium sulfate do you need to make 196 mL of a 0.317 M solution?

Break down the given concentration into its corresponding *PER* expression. Be careful with the details.

0.317 mol K_2SO_4/L solution

Construct the *PLAN* for the problem.

GIVEN: 196 mL; 0.317 mol K_2SO_4/L solution *WANTED*: g K_2SO_4

PER/PATH: mL $\xrightarrow{\text{1000 mL/L}}$ L $\xrightarrow{\text{0.317 mol } K_2SO_4\text{/L}}$ mol K_2SO_4 $\xrightarrow{\text{174.27 g } K_2SO_4\text{/mol } K_2SO_4}$ g K_2SO_4

Execute your *PLAN*. Find the final answer.

$$196 \text{ mL} \times \frac{1 \text{ L}}{1000 \text{ mL}} \times \frac{0.317 \text{ mol } K_2SO_4}{\text{L}} \times \frac{174.27 \text{ g } K_2SO_4}{\text{mol } K_2SO_4} = 10.8 \text{ g } K_2SO_4$$

EXAMPLE 16.5-WB

What volume (mL) of 0.073 M calcium nitrate solution is required for an experimental procedure that calls for 1.0 gram of calcium ion?

The solution being used as a source of calcium ion is 0.073 M calcium nitrate. What is the molar concentration of calcium ion in this solution? Explain how you arrived at your answer.

0.073 M Ca^{2+}

Since $Ca(NO_3)_2(aq) \rightarrow Ca^{2+}(aq) + 2\ NO_3^-(aq)$, there is one mole of calcium ion per one mole of calcium nitrate.

PLAN the solution. Don't forget that the question asks for volume in milliliters, the most common laboratory unit for measuring volume of liquids.

GIVEN: 1.0 g Ca^{2+}; 0.073 M Ca^{2+} *WANTED:* mL

PER/PATH: g Ca^{2+} $\xrightarrow{40.08\ g\ Ca^{2+}/mol\ Ca^{2+}}$ mol Ca^{2+} $\xrightarrow{0.073\ mol\ Ca^{2+}/L}$ L $\xrightarrow{1000\ mL/L}$ mL

Determine the final answer.

$$1.0\ g\ Ca^{2+} \times \frac{1\ mol\ Ca^{2+}}{40.08\ g\ Ca^{2+}} \times \frac{1\ L}{0.073\ mol\ Ca^{2+}} \times \frac{1000\ mL}{L} = 3.4 \times 10^2\ mL$$

EXAMPLE 16.6-WB

If 3.14 g CHI_3 is dissolved in 23.52 g of solvent, what is the molality of the CHI_3 in this solution?

Write the definition of molality so that you can clearly see the two quantities that you need to find.

$$m \equiv \frac{mol\ solute}{kg\ solvent}$$

We can find each part of the molality fraction separately and then substitute into the definition to finish the problem. *PLAN* your approach to find moles of solute, then carry out the calculation.

GIVEN: 3.14 g CHI_3 *WANTED*: mol CHI_3

PER/PATH: g CHI_3 $\xrightarrow{393.7\ g\ CHI_3/mol\ CHI_3}$ mol CHI_3

$$3.14\ g\ CHI_3 \times \frac{1\ mol\ CHI_3}{393.7\ g\ CHI_3} = 7.98 \times 10^{-3}\ mol\ CHI_3$$

Now find the value of the denominator, kilograms of solvent. *PLAN* and solve.

GIVEN: 23.52 g solvent *WANTED*: kg solvent

PER/PATH: g solvent $\xrightarrow{\text{1000 g solvent/kg solvent}}$ kg solvent

$$23.52 \text{ g solvent} \times \frac{1 \text{ kg solvent}}{1000 \text{ g solvent}} = 0.02352 \text{ kg solvent}$$

You now have each part of the molality ratio. Find the final answer.

GIVEN: 7.98×10^{-3} mol CHI_3; 0.02352 kg solvent *WANTED*: m

EQUATION: $m \equiv \frac{\text{mol solute}}{\text{kg solvent}} = \frac{7.98 \times 10^{-3} \text{ mol } CHI_3}{0.02352 \text{ kg solvent}} = 0.339 \text{ m } CHI_3$

EXAMPLE 16.7-WB

How many milliliters of water should be used to dissolve 13.5 g $C_6H_{12}O_6$ in preparing a 0.255 molal solution?

The question assumes that you know the density of water. Other than that, this problem is similar to Example 16.7 in the textbook, which you just solved. Go all the way to the final answer.

GIVEN: 13.5 g $C_6H_{12}O_6$ *WANTED:* mL H_2O

PER/PATH: g $C_6H_{12}O_6$ $\xrightarrow{180.16\text{ g } C_6H_{12}O_6/\text{mol } C_6H_{12}O_6}$ mol $C_6H_{12}O_6$

$\xrightarrow{0.255\text{ mol } C_6H_{12}O_6/\text{kg } H_2O}$ kg H_2O $\xrightarrow{1000\text{ g } H_2O/\text{kg } H_2O}$ g H_2O

$\xrightarrow{1.00\text{ g } H_2O/\text{mL } H_2O}$ mL H_2O

$$13.5\text{ g } C_6H_{12}O_6 \times \frac{1\text{ mol } C_6H_{12}O_6}{180.16\text{ g } C_6H_{12}O_6} \times \frac{1\text{ kg } H_2O}{0.255\text{ mol } C_6H_{12}O_6} \times \frac{1000\text{ g } H_2O}{\text{kg } H_2O} \times \frac{1.00\text{ mL } H_2O}{\text{g } H_2O} = 294\text{ mL } H_2O$$

EXAMPLE 16.8-WB

State the number of equivalents of acid and base per mole in each of the following reactions:

	eq acid/mol	eq base/mol
$H_2SO_4 + KOH \rightarrow KHSO_4 + H_2O$		
$H_2SO_4 + 2\ KOH \rightarrow K_2SO_4 + 2\ H_2O$		
$H_3AsO_4 + NaOH \rightarrow KH_2AsO_4 + H_2O$		
$H_3AsO_4 + Ca(OH)_2 \rightarrow CaHAsO_4 + 2\ H_2O$		

Use the columns provided to write your answers.

---

	eq acid/mol	eq base/mol
$H_2SO_4 + KOH \rightarrow KHSO_4 + H_2O$	1	1
$H_2SO_4 + 2\ KOH \rightarrow K_2SO_4 + 2\ H_2O$	2	1
$H_3AsO_4 + NaOH \rightarrow KH_2AsO_4 + H_2O$	1	1
$H_3AsO_4 + Ca(OH)_2 \rightarrow CaHAsO_4 + 2\ H_2O$	2	2

Sulfuric acid, H_2SO_4, can provide either one or two moles of H^+ per mole of H_2SO_4. In the first reaction, it reacts with one mole of OH^-, and in the second reaction, it reacts with two moles of OH^-. Arsenic acid, H_3AsO_4, can provide one, two, or three moles of H^+. In the third reaction, it reacts with one mole of OH^-, and in the fourth reaction, it reacts with two moles of OH^-.

EXAMPLE 16.9-WB

State the number of equivalents of acid per mole H_3PO_4 and the equivalent mass of H_3PO_4 in the following reaction: $2\ H_3PO_4(aq) + 3\ Ca(OH)_2(s) \rightarrow Ca_3(PO_4)_2(s) + 6\ H_2O(\ell)$.

Determine the number of equivalents of acid, calculate the molar mass, and then find the equivalent mass.

All three hydrogen ions from H_3PO_4 react; there are 3 eq acid/mol H_3PO_4.

$$\frac{97.99 \text{ g } H_3PO_4}{3 \text{ eq } H_3PO_4} = 32.7 \text{ g } H_3PO_4/\text{eq } H_3PO_4$$

EXAMPLE 16.10-WB

How many equivalents are in 10.0 grams of potassium hydrogen phthalate in the reaction $KHC_8H_4O_4 \rightarrow K^+ + H^+ + C_8H_4O_4^{2-}$?

Once you've figured out the number of equivalents per mole of acid and the molar mass, you have the equivalent mass ratio. It's a one-step conversion from grams to equivalents. Take it all the way.

GIVEN: 10.0 g $KHC_8H_4O_4$; 204.22 g/mol $KHC_8H_4O_4$; 1 eq/mol $KHC_8H_4O_4$ *WANTED:* eq

PER/PATH: g $KHC_8H_4O_4$ $\xrightarrow{204.22 \text{ g } KHC_8H_4O_4/1 \text{ eq}}$ eq

$$10.0 \text{ g } KHC_8H_4O_4 \times \frac{1 \text{ eq}}{204.22 \text{ g } KHC_8H_4O_4} = 0.0490 \text{ eq}$$

EXAMPLE 16.11-WB

2.50×10^2 mL of a sulfuric acid solution contains 10.5 g H_2SO_4. Calculate its normality for the reaction $H_2SO_4 + 2\ NaOH \rightarrow Na_2SO_4 + 2\ H_2O$.

The procedure is the same here as in textbook Example 16.11. Be careful with the number of equivalents per mole. Solve the problem completely.

GIVEN: 10.5 g H_2SO_4; 2.50×10^2 mL *WANTED*: N (eq/L)

PER/PATH: g H_2SO_4 $\xrightarrow{98.09 \text{ g } H_2SO_4/2 \text{ eq } H_2SO_4}$ eq H_2SO_4

$$10.5 \text{ g } H_2SO_4 \times \frac{2 \text{ eq } H_2SO_4}{98.09 \text{ g } H_2SO_4} = 0.214 \text{ eq } H_2SO_4$$

EQUATION: $N \equiv \frac{eq}{L} = \frac{0.214 \text{ eq } H_2SO_4}{2.50 \times 10^2 \text{ mL}} \times \frac{1000 \text{ mL}}{L} = 0.856 \text{ eq } H_2SO_4/L = 0.856 \text{ N } H_2SO_4$

Both hydrogens are released from H_2SO_4 in this reaction, so there are two equivalents per mole. Again, milliliters must be converted to liters to be used in the defining equation for normality.

EXAMPLE 16.12-WB

What volume of 0.225 N hydrochloric acid should be measured for a reaction that requires 0.077 equivalents of acid?

Write your *PLAN* and solve.

GIVEN: 0.077 eq; 0.225 N HCl *WANTED:* volume HCl (assume mL)

PER/PATH: eq $\xrightarrow{0.225 \text{ eq/L}}$ L $\xrightarrow{1000 \text{ mL/L}}$ mL

$$0.077 \text{ eq} \times \frac{1 \text{ L}}{0.225 \text{ eq}} \times \frac{1000 \text{ mL}}{L} = 3.4 \times 10^2 \text{ mL}$$

EXAMPLE 16.13-WB

How many grams of potassium hydroxide are needed to make 855 mL of a 0.716 N solution?

Take it all the way to the answer.

GIVEN: 855 mL *WANTED*: g KOH

PER/PATH: mL $\xrightarrow{1000\ \text{mL/L}}$ L $\xrightarrow{0.716\ \text{eq KOH/L}}$ eq KOH $\xrightarrow{56.11\ \text{g KOH/eq KOH}}$ g KOH

$$855\ \text{mL} \times \frac{1\ \text{L}}{1000\ \text{mL}} \times \frac{0.716\ \text{eq KOH}}{\text{L}} \times \frac{56.11\ \text{g KOH}}{\text{eq KOH}} = 34.3\ \text{g KOH}$$

KOH can react with one mol of hydrogen ions, so there is 1 eq KOH/mol KOH.

EXAMPLE 16.14-WB

Concentrated acetic acid is 17.5 M; vinegar is 0.87 M acetic acid. How much concentrated acid do you need to make 32 L of vinegar?

Carefully assign values to each variable (V_c, M_c, V_d, M_d), and then solve the dilution equation for the unknown and calculate the answer.

GIVEN: M_c = 17.5 M; M_d = 0.87 M; V_d = 32 L *WANTED*: V_c (assume L)

EQUATION: $V_c = \frac{V_d \times M_d}{M_c} = \frac{32\ \text{L} \times 0.87\ \text{M}}{17.5\ \text{M}} = 1.6\ \text{L}$

We assumed that the volume should be expressed in liters instead of the more common laboratory unit milliliters because of the relatively large volume of vinegar.

EXAMPLE 16.15-WB

A laboratory technician adds 20.0 mL of 15.8 M nitric acid to 2.00 L of water. What is the concentration of the resulting solution?

What is the volume of the diluted solution?

$$2.00\ \text{L} \times \frac{1000\ \text{mL}}{\text{L}} + 20.0\ \text{mL} = 2.02 \times 10^3\ \text{mL}$$

Note that the volume of water, when expressed in milliliters, has its doubtful digit in the tens column. This limits the total volume to the tens column. You may have chosen to express the volumes in liters, which is equally acceptable, as long as both volume units are the same. In this case, V_d = 2.02 L.

Take it the rest of the way.

GIVEN: $V_c = 20.0$ mL; $M_c = 15.8$ M; $V_d = 2.02 \times 10^3$ mL *WANTED:* M_d

EQUATION: $M_d = \dfrac{V_c \times M_c}{V_d} = \dfrac{20.0 \text{ mL} \times 15.8 \text{ M}}{2.02 \times 10^3 \text{ mL}} = 0.156 \text{ M}$

EXAMPLE 16.16-WB

Find the number of milliliters of 0.103 M potassium chloride required to precipitate as lead(II) chloride all the lead in a solution that contains 0.293 gram of lead(II) nitrate.

Begin with the balanced equation, *GIVEN, WANTED,* and *PER/PATH.*

$2\ KCl + Pb(NO_3)_2 \rightarrow 2\ KNO_3 + PbCl_2$

GIVEN: 0.293 g $Pb(NO_3)_2$ *WANTED*: mL KCl

PER/PATH: g $Pb(NO_3)_2$ $\xrightarrow{331.2 \text{ g } Pb(NO_3)_2/\text{mol } Pb(NO_3)_2}$ mol $Pb(NO_3)_2$

$\xrightarrow{2 \text{ mol KCl}/1 \text{ mol } Pb(NO_3)_2}$ mol KCl $\xrightarrow{0.103 \text{ mol KCl}/1000 \text{ mL KCl}}$ mL KCl

We used the one step moles-to-milliliters conversion at the end of the *PER/PATH.* There is no problem with converting from moles to liters followed by a liters to milliliters conversion, if you so desire.

Complete the problem.

$$0.293 \text{ g } Pb(NO_3)_2 \times \frac{1 \text{ mol } Pb(NO_3)_2}{331.2 \text{ g } Pb(NO_3)_2} \times \frac{2 \text{ mol KCl}}{1 \text{ mol } Pb(NO_3)_2} \times \frac{1000 \text{ mL KCl}}{0.103 \text{ mol KCl}} = 17.2 \text{ mL KCl}$$

EXAMPLE 16.17-WB

How many liters of hydrogen, measured at STP, will be released by the complete reaction of 45.0 mL 0.486 M H_2SO_4 with excess solid zinc?

This example asks you to calculate the "volume of wanted gas at given temperature and pressure" (from Equation 16.12). If you have studied gas stoichiometry, you probably recall the conversion factor between moles of a gas and its volume at STP (standard temperature and pressure). It is one of the numbers "worth remembering" from Section 9.3, 13.6, or 14.5. Write this *PER* expression. (If you have not studied gas stoichiometry, skip to the next step.)

22.4 L H_2/mol H_2

This *PER* expression can be applied when the unit path reaches moles of H_2. From there, you can do a one-step conversion to volume of H_2 in liters.

PLAN, set up, and solve the problem.

$Zn(s) + H_2SO_4(aq) \rightarrow ZnSO_4(aq) + H_2(g)$

GIVEN: 45.0 mL H_2SO_4 *WANTED*: L H_2

PER/PATH: mL H_2SO_4 $\xrightarrow{1000 \text{ mL } H_2SO_4/\text{L } H_2SO_4}$ L H_2SO_4 $\xrightarrow{0.486 \text{ mol } H_2SO_4/\text{L } H_2SO_4}$

mol H_2SO_4 $\xrightarrow{1 \text{ mol } H_2/1 \text{ mol } H_2SO_4}$ mol H_2 $\xrightarrow{22.4 \text{ L } H_2/\text{mol } H_2}$ L H_2

$$45.0 \text{ mL } H_2SO_4 \times \frac{1 \text{ L } H_2SO_4}{1000 \text{ mL } H_2SO_4} \times \frac{0.486 \text{ mol } H_2SO_4}{\text{L } H_2SO_4} \times \frac{1 \text{ mol } H_2}{1 \text{ mol } H_2SO_4} \times \frac{22.4 \text{ L } H_2}{\text{mol } H_2} = 0.490 \text{ L } H_2$$

EXAMPLE 16.18-WB

A student plans to carry out a reaction using solutions of calcium nitrate and sodium carbonate, intending to form a precipitate of calcium carbonate. He has 100.0 mL of 0.42 M calcium nitrate and a 10.0-L bottle of 0.58 M sodium carbonate. How much of the sodium carbonate solution should he use if he wants to add just enough to get the maximum amount of solid precipitate?

The question is asking for the volume of sodium carbonate solution needed to react with 100.0 mL of 0.42 M calcium nitrate solution. The fact that there is 10.0-L of sodium carbonate available only tells you that there is more than enough of that solution to complete the reaction. Assume that the final answer should be expressed in milliliters. Using millimoles and milliliters, *PLAN* the solution, including the reaction equation.

$Ca(NO_3)_2(aq) + Na_2CO_3(aq) \rightarrow CaCO_3(s) + 2\ NaNO_3(aq)$

GIVEN: 100.0 mL 0.42 M $Ca(NO_3)_2$ *WANTED:* mL 0.58 M Na_2CO_3

PER/PATH: mL $Ca(NO_3)_2$ $\xrightarrow{0.42\text{ mmol }Ca(NO_3)_2/\text{mL }Ca(NO_3)_2}$ mmol $Ca(NO_3)_2$

$\xrightarrow{1\text{ mmol }Na_2CO_3/1\text{ mmol }Ca(NO_3)_2}$ mmol Na_2CO_3 $\xrightarrow{0.58\text{ mmol }Na_2CO_3/\text{mL }Na_2CO_3}$ mL Na_2CO_3

Complete the setup and determine the final answer.

$$100.0\text{ mL }Ca(NO_3)_2 \times \frac{0.42\text{ mmol }Ca(NO_3)_2}{\text{mL }Ca(NO_3)_2} \times \frac{1\text{ mmol }Na_2CO_3}{1\text{ mmol }Ca(NO_3)_2} \times \frac{1\text{ mL }Na_2CO_3}{0.58\text{ mmol }Na_2CO_3} = 72.4\text{ mL }Na_2CO_3$$

EXAMPLE 16.19-WB

$HSO_3NH_2(aq) + KOH(aq) \rightarrow H_2O(\ell) + KSO_3NH_2(aq)$ is the equation for the reaction by which a potassium hydroxide solution is standardized by titrating against a weighed quantity of sulfamic acid. A laboratory technician uses 34.2 mL of the solution to neutralize 0.395 gram of HSO_3NH_2. Find the molarity of the KOH.

The first part of the problem requires you to convert from grams of HSO_3NH_2 to moles of KOH. The balanced equation is already provided for you. *PLAN* and solve.

GIVEN: 0.395 g HSO_3NH_2 *WANTED:* mol KOH

PER/PATH: g HSO_3NH_2 $\xrightarrow{97.10\text{ g }HSO_3NH_2/\text{mol }HSO_3NH_2}$ mol HSO_3NH_2 $\xrightarrow{1\text{ mol KOH}/1\text{ mol }HSO_3NH_2}$ mol KOH

$$0.395\text{ g }HSO_3NH_2 \times \frac{1\text{ mol }HSO_3NH_2}{97.10\text{ g }HSO_3NH_2} \times \frac{1\text{ mol KOH}}{1\text{ mol }HSO_3NH_2} = 0.00407\text{ mol KOH}$$

Now that you know both the number of moles of KOH and the volume of KOH, you can find the molarity.

EQUATION: $M \equiv \frac{\text{mol}}{\text{L}} = \frac{0.00407\text{ mol}}{34.2\text{ mL}} \times \frac{1000\text{ mL}}{\text{L}} = 0.119\text{ M KOH}$

EXAMPLE 16.20-WB

Find the molarity of a solution of sodium hydroxide if 21.4 mL are required to react with 20.0 mL 0.101 M hydrochloric acid in a titration experiment.

Go for the complete solution.

$NaOH + HCl \rightarrow NaCl + H_2O$

GIVEN: 21.4 mL NaOH; 20.0 mL 0.101 M HCl *WANTED*: M NaOH

PER/PATH: mL HCl $\xrightarrow{\text{0.101 mol HCl/1000 mL HCl}}$ mol HCl $\xrightarrow{\text{1 mol NaOH/1 mol HCl}}$ mol NaOH

$$20.0 \text{ mL HCl} \times \frac{0.101 \text{ mol HCl}}{1000 \text{ mL HCl}} \times \frac{1 \text{ mol NaOH}}{1 \text{ mol HCl}} = 0.00202 \text{ mol NaOH}$$

EQUATION: $M \equiv \frac{\text{mol}}{\text{L}} = \frac{0.00202 \text{ mol NaOH}}{21.4 \text{ mL NaOH}} \times \frac{1000 \text{ mL NaOH}}{\text{L NaOH}} = 0.0944 \text{ M NaOH}$

EXAMPLE 16.21-WB

Exactly 22.73 mL of sulfuric acid is needed to titrate 0.504 g of sodium carbonate in the reaction $H_2SO_4(aq) + Na_2CO_3(aq) \rightarrow CO_2(g) + Na_2SO_4(aq) + H_2O(\ell)$ What is the normality of the sulfuric acid in this reaction?

Find the number of equivalents of sulfuric acid.

GIVEN: 22.73 mL H_2SO_4; 0.504 g Na_2CO_3 *WANTED*: eq H_2SO_4

PER/PATH: g Na_2CO_3 $\xrightarrow{105.99 \text{ g } Na_2CO_3/2 \text{ eq } Na_2CO_3}$ eq Na_2CO_3

$$0.504 \text{ g } Na_2CO_3 \times \frac{2 \text{ eq } H_2SO_4 \text{ or } Na_2CO_3}{105.99 \text{ g } Na_2CO_3} = 0.00951 \text{ eq } H_2SO_4$$

Complete the problem by substitution of the given or known numerator and denominator into the definition of normality.

EQUATION: $N \equiv \frac{\text{eq}}{\text{L}} = \frac{0.00951 \text{ eq } H_2SO_4}{22.73 \text{ mL}} \times \frac{1000 \text{ mL}}{\text{L}} = 0.0.418 \text{ N } H_2SO_4$

EXAMPLE 16.22-WB

Find the normality of a solution of sodium hydroxide if 23.7 mL is required to react with 25.0 mL of 0.236 N phosphoric acid in the reaction 3 NaOH(aq) + H_3PO_4(aq) → Na_3PO_4(aq) + 3 $H_2O(\ell)$.

Complete the example.

GIVEN: V_1 = 25.0 mL; N_1 = 0.236 N; V_2 = 23.7 mL *WANTED*: N_2

EQUATION: $N_2 = \frac{V_1N_1}{V_2} = \frac{25.0 \text{ mL} \times 0.236 \text{ N}}{23.7 \text{ mL}} = 0.249 \text{ N NaOH}$

EXAMPLE 16.23-WB

A certain solvent freezes at 84.4°C. If 12.2 g of naphthalene, $C_{10}H_8$, is dissolved in 135 g of the solvent, the freezing temperature is 81.8°C. Calculate K_f for the solvent.

Write the definition of K_f so that you can see the two quantities that make up its ratio.

$$K_f \equiv \frac{\Delta T_f}{m}$$

You need to find both ΔT_f and m, the molality of the solution. Begin with the ΔT_f calculation.

$$\Delta T_f \equiv T_f - T_i = 81.8°C - 84.4°C = -2.6°C$$

The freezing point *depression* is 2.6°C. Recall that we treat this quantity as a positive number in calculations because the term "depression" clearly identifies the direction of the temperature change.

Now you need the molality of the solution. Write its definition to clearly see the needed quantities, and then go all the way to the value of molality.

$$m \equiv \frac{\text{mol solute}}{\text{kg solvent}}$$

GIVEN: 12.2 g $C_{10}H_8$ *WANTED*: mol $C_{10}H_8$

PER/PATH: g $C_{10}H_8$ $\xrightarrow{128.16 \text{ g } C_{10}H_8/\text{mol } C_{10}H_8}$ mol $C_{10}H_8$

$$12.2 \text{ g } C_{10}H_8 \times \frac{1 \text{ mol } C_{10}H_8}{128.16 \text{ g } C_{10}H_8} = 0.0952 \text{ mol } C_{10}H_8$$

GIVEN: 135 g solvent *WANTED:* kg solvent

PER/PATH: g solvent $\xrightarrow{1000 \text{ g solvent/kg solvent}}$ kg solvent

$$135 \text{ g solvent} \times \frac{1 \text{ kg solvent}}{1000 \text{ g solvent}} = 0.135 \text{ g solvent}$$

$$m \equiv \frac{\text{mol solute}}{\text{kg solvent}} = \frac{0.0952 \text{ mol } C_{10}H_8}{0.135 \text{ kg solvent}} = 0.705 \text{ m } C_{10}H_8$$

Now that you've found both parts in the definition of K_f, you can complete the problem with the final setup and calculation.

GIVEN: $\Delta T_f = 2.6°C$ *WANTED:* m = 0.705 m

EQUATION: $K_f = \frac{\Delta T_f}{m} = \frac{2.6°C}{0.705 \text{ m}} = 3.7°C/m$

EXAMPLE 16.24-WB

The molal freezing point constant for water is 1.86°C/m. If 3.52 g of an unknown is dissolved in 20.79 g water, the resultant solution freezes at –1.97°C, as measured with a thermometer that gives the freezing point of pure water as 0.00°C. What is the molar mass of this unknown?

Write the definition of molar mass.

$$MM \equiv \frac{g}{mol}$$

This is the final goal for this problem, the molar mass of the solute.

The numerator of the molar mass fraction, grams of solute, is given in the problem statement as 3.52 g. Thus the denominator is the quantity that needs to be determined. The problem statement gives freezing point depression information that will lead to the molality of the solution. This, in turn, can be used along with the mass of the solvent to find moles.

Use ΔT_f and K_f to find the molality of the solution.

GIVEN: $\Delta T_f = 1.97°C$; $K_f = 1.86°C/m$ *WANTED*: m (mol solute/kg H_2O)

EQUATION: $m = \frac{\Delta T_b}{K_b} = 1.97°C \times \frac{m}{1.86°C} = 1.06\ m = \frac{1.06 \text{ mol solute}}{\text{kg } H_2O}$

Now you can use the mass of solvent and the molality to find the moles of solute.

GIVEN: 20.79 g H_2O; 1.06 mol solute/kg H_2O *WANTED*: mol solute

PER/PATH: g H_2O $\xrightarrow{1000 \text{ g } H_2O/\text{kg } H_2O}$ kg H_2O $\xrightarrow{1.06 \text{ mol solute/kg } H_2O}$ mol solute

$$20.79 \text{ g } H_2O \times \frac{1 \text{ kg } H_2O}{1000 \text{ g } H_2O} \times \frac{1.06 \text{ mol solute}}{\text{kg } H_2O} = 0.0220 \text{ mol solute}$$

Both parts of the molar mass ratio are now known. Complete the problem.

GIVEN: 0.0220 mol solute; 3.52 g solute *WANTED*: MM

EQUATION: $MM \equiv \frac{g}{mol} = \frac{3.52 \text{ g}}{0.0220 \text{ mol}} = 1.60 \times 10^2 \text{ g/mol}$

Name: ______________________ Date: ______________________

ID: ______________________ Section: ______________________

Questions, Exercises, and Problems

Section 16.1: The Characteristics of a Solution

1. Mixtures of gases are always true solutions. True or false? Explain why.

2. Can you see particles in a solution? If yes, give an example.

Section 16.2: Solution Terminology

3. Distinguish between the solute and solvent in each of the following solutions: (a) saltwater [NaCl(aq)]; (b) sterling silver (92.5% Ag, 7.5% Cu); (c) air (about 80% N_2, 20% O_2). On what do you base your distinctions?

4. Would it be proper to say that a saturated solution is a concentrated solution? or that a concentrated solution is a saturated solution? Point out the distinctions between these sometimes confused terms.

Name: ______________________ Date: ______________________

ID: ______________________ Section: ______________________

5. What happens if you add a very small amount of solid salt (NaCl) to each of the beakers described below? Include a statement about the *amount* of solid eventually found in the beaker, compare with the amount you added: (a) a beaker containing *saturated* NaCl solution, (b) a beaker with *unsaturated* NaCl solution, (c) a beaker containing *supersaturated* NaCl solution.

6. In stating solubility, an important variable must be specified. What is that variable, and how does solubility of a solid solute usually depend on it?

7. When acetic acid, a clear, colorless liquid, and water are mixed, a clear, uniform colorless liquid results. Is acetic acid soluble in water? Is acetic acid miscible in water? Explain your answers. If you answers are different from those to textbook Question 7, explain why they are different.

Section 16.3: The Formation of a Solution

8. How is it that both cations and anions, positively charged ions and negatively charged ions, can be hydrated by the same substance, water?

9. Explain why the dissolving process is reversible.

Name: ______________________ Date: ______________________

ID: ______________________ Section: ______________________

10. Describe the changes that occur between the time excess solute is placed into water and the time the solution becomes saturated.

11. At what time during the development of a saturated solution is the rate at which ions move from solvent to solute [solute(aq) → solute(s)] greater than the rate from solute to solvent [solute(s) → solute(aq)]?

12. Why can you not prepare a supersaturated solution by adding more solute and stirring until it dissolved?

13. Explain how each of the acts in textbook Question 13 speeds the dissolving process.

Section 16.4: Factors that Determine Solubility

Questions 14 and 15: Structural diagrams for several substances are given in Table 16.2 on Page 467 of the textbook.

14. Which of the following solutes do you expect to be more soluble in water than in cyclohexane: (a) formic acid, (b) benzene, (c) methylamine, (d) tetrafluoromethane? Explain your choice(s).

Name: ______________________ Date: ______________________

ID: ______________________ Section: ______________________

15. Which compound, glycerine or hexane, do you expect would be more miscible in water? Why?

16. On opening a bottle of carbonated beverage, many bubbles are released and the sound of escaping gas is heard. This suggests that the beverage is bottled under high pressure. Yet, for safety reasons, the pressure cannot be much more than one atmosphere. What gas do you suppose is in the small space between the beverage and the cap of a bottle of carbonated beverage before the cap is removed?

Section 16.5: Solution Concentration: Percentage by Mass

17. Calculate the percentage concentration of a solution prepared by dissolving 2.32 g of calcium chloride in 81.0 g of water.

Answer: [ ]

18. How many grams of ammonium nitrate must be weighed out to make 415 g of a 58.0% solution? In how many milliliters of water should it be dissolved?

Answer: [ ]

Name: ______________________ Date: ______________________

ID: ______________________ Section: ______________________

Section 16.6: Solution Concentration: Molarity

19. Potassium iodide is the additive in "iodized" table salt. Calculate the molarity of a solution prepared by dissolving 2.41 g of potassium iodide in water and diluting to 50.0 mL.

Answer: [ ]

20. 18.0 g of anhydrous nickel chloride is dissolved in water and diluted to 90.0 mL. 30.0 g of nickel chloride hexahydrate are also dissolved in water and diluted to 90.0 mL. Identify the solution with higher molar concentration and calculate its molarity.

Answer: [ ]

21. Large quantities of silver nitrate are used in making photographic chemicals. Find the mass that must be used in preparing 2.50×10^2 mL 0.058 M silver nitrate.

Answer: [ ]

Name: ______________________ Date: ______________________

ID: ______________________ Section: ______________________

22. Potassium hydroxide is used in making liquid soap, as well as many other things. How many grams would you use to prepare 2.50 L of 1.40 M potassium hydroxide?

Answer: [ ]

23. What volume of concentrated sulfuric acid, which is 18 molar, is required to obtain 5.19 mole of the acid?

Answer: [ ]

24. 0.132 M sodium chloride is to be the source of 8.33 g of dissolved solute. What volume of solution is needed?

Answer: [ ]

Name: ____________________ Date: ____________________

ID: ____________________ Section: ____________________

25. Calculate the moles of silver nitrate in 55.7 mL 0.204 M silver nitrate.

Answer: [ ]

26. Despite its intense purple color, potassium permanganate is used in bleaching operations. How many moles are in 25.0 mL of 0.0841 M $KMnO_4$?

Answer: [ ]

27. The density of 3.30 M KNO_3 is 1.15 g/mL. What is its percentage concentration?

Answer: [ ]

Name: ______________________ Date: ______________________

ID: ______________________ Section: ______________________

Section 16.7: Solution Concentration: Molality (Optional)

28. Calculate the molal concentration of a solution of 44.9 g of naphthalene, $C_{10}H_8$, in 175 g of benzene, C_6H_6.

Answer: [ ]

29. Diethylamine, $(CH_3CH_2)NH$, is highly soluble in ethanol, C_2H_5OH. Calculate the number of grams of diethylamine that would be dissolved in 4.00×10^2 g of ethanol to produce 4.70 m $(CH_3CH_2)NH$.

Answer: [ ]

30. How many milliliters of water are needed to dissolve 97.7 mg sodium chloride in the preparation of a 2.80×10^{-3} m solution?

Answer: [ ]

Name: ______________________ Date: ______________________

ID: ______________________ Section: ______________________

Section 16.8: Solution Concentration: Normality (Optional)

31. What is equivalent mass? Why can you state positively the equivalent mass of LiOH, but not H_2SO_4?

32. State the number of equivalents in one mole of HNO_2; in one mole of H_2SeO_4 in $H_2SeO_4 \rightarrow H^+ + HSeO_4^-$. (Se is selenium, Z = 34.)

33. State the maximum number of equivalents per mole of $Cu(OH)_2$, per mole of $Fe(OH)_3$.

34. Calculate the equivalent masses of HNO_2 and H_2SeO_4 in Question 32.

35. What are the equivalent masses of $Cu(OH)_2$ and $Fe(OH)_3$ in Question 33?

Name: ______________________ Date: ______________________

ID: ______________________ Section: ______________________

36. What is the normality of the solution made when 2.25 g potassium hydroxide are dissolved in water and diluted to 2.50×10^2 mL?

Answer: [ ]

37. 6.69 g $H_2C_2O_4$ is dissolved in water, diluted to 2.00×10^2 mL, and used in a reaction in which it ionizes as follows: $H_2C_2O_4 \rightarrow H^+ + HC_2O_4^-$. What is the normality of the solution?

Answer: [ ]

38. $NaHSO_4$ is used as an acid in the reaction $HSO_4^- \rightarrow H^+ + SO_4^{2-}$. What mass of $NaHSO_4$ must be dissolved in 7.50×10^2 mL of solution to produce 0.200 N $NaHSO_4$?

Answer: [ ]

Name: ____________________ Date: ____________________

ID: ____________________ Section: ____________________

39. What is the molarity of (a) 0.965 N sodium hydroxide, (b) 0.237 N H_3PO_4 in $H_3PO_4 + 2\ NaOH \rightarrow Na_2HPO_4 + 2\ H_2O$?

40. 73.1 mL 0.834 N NaOH has how many equivalents of solute?

Answer: ____________________

41. What volume of 0.492 N $KMnO_4$ contains 0.788 eq?

Answer: ____________________

Section 16.10: Dilution of Concentrated Solutions

42. What is the molarity of the acetic acid solution if 45.0 mL of 17 M $HC_2H_3O_2$ is diluted to 1.5 L?

Answer: ____________________

Name: ______________________ Date: ______________________

ID: ______________________ Section: ______________________

43. How many milliliters of concentrated nitric acid, 16 M HNO_3, will you use to prepare 7.50×10^2 mL of 0.69 M HNO_3?

Answer: [ ]

44. Calculate the volume of 18 M H_2SO_4 required to prepare 3.0 L of 2.9 N H_2SO_4 for the reactions in which the sulfuric acid is completely ionized.

Answer: [ ]

45. Calculate the normality of a solution prepared by diluting 15.0 mL of 15 M H_3PO_4 to 2.50×10^2 mL. The solution will be used in the same reaction as that in workbook Question 39.

Answer: [ ]

Name: ______________________ Date: ______________________

ID: ______________________ Section: ______________________

Section 16.11: Solution Stoichiometry

46. Calculate the grams of magnesium hydroxide that will precipitate from 25.0 mL of 0.398 M magnesium chloride by the addition of excess sodium hydroxide solution.

Answer: [ ]

47. Calculate the mass of calcium phosphate that will precipitate when excess sodium phosphate solution is added to 100.0 mL of 0.130 M calcium nitrate.

Answer: [ ]

48. How many milliliters of 1.50 M NaOH must react with aluminum to yield 2.00 L of hydrogen, measured at 22°C and 789 torr, by the reaction $2\ Al + 6\ NaOH \rightarrow 2\ Na_3AlO_3 + 3\ H_2$? Assume complete conversion of reactants to products.

Answer: [ ]

Name: ______________________ Date: ______________________

ID: ______________________ Section: ______________________

49. What volume of 0.842 M NaOH would react with 8.74 g of sulfamic acid, NH_2SO_3H, a solid acid with one replaceable hydrogen?

Answer: [ ]

Section 16.12: Titration Using Molarity

50. $2\ HC_7H_5O_2 + Na_2CO_3 \rightarrow 2\ NaC_7H_5O_2 + H_2O + CO_2$ is the equation for a reaction by which a solution of sodium carbonate may be standardized. 5.038 g of $HC_7H_5O_2$ uses 51.89 mL of the sodium carbonate solution in the titration. Find the molarity of the sodium carbonate.

Answer: [ ]

51. A student is to titrate solid maleic acid, $H_2C_4H_2O_4$ (two replaceable hydrogens) with a KOH solution of unknown concentration. She dissolves 1.45 g of maleic acid in water and titrates 50.0 mL of the base to neutralize the acid. What is the molarity of the KOH solution?

Answer: [ ]

Name: ______________________ Date: ______________________

ID: ______________________ Section: ______________________

52. 37.80 mL of a 0.4052 M $NaHCO_3$ solution are required to titrate a 20.00 mL sample of sulfuric acid solution. What is the molarity of the acid? The reaction equation is $H_2SO_4 + 2\ NaHCO_3 \rightarrow Na_2SO_4 + 2\ H_2O + 2\ CO_2$.

Answer: [ ]

Section 16.13: Titration Using Normality (Optional)

53. What is the normality of the sodium carbonate solution in Question 50?

Answer: [ ]

54. What is the normality of the base in Question 51?

Answer: [ ]

Name: ______________________ Date: ______________________

ID: ______________________ Section: ______________________

55. Calculate the normality of a solution of sodium carbonate if a 25.0 mL sample requires 39.8 mL of 0.405 N sulfuric acid in a titration.

Answer: [ ]

56. 42.2 mL 0.402 N sodium hydroxide is required to titrate 50.0 mL of a solution of tartaric acid ($H_2C_4H_4O_6$) of unknown concentration. Find the normality of the acid.

Answer: [ ]

57. Repeating the titration described in textbook Question 57, but with a different indicator, it is found that only 16.3 mL 0.208 N sodium hydroxide is required for 20.0 mL of the phosphoric acid solution. Calculate the normality of the acid and account for the difference in the answers in textbook Question 57 and this question.

Answer: [ ]

58. 1.21 g of an organic compound that functions as a base in reaction with sulfuric acid is dissolved in water and titrated with 0.170 N sulfuric acid. What is the equivalent mass of the base if 30.7 mL of acid is required in the titration?

Answer: [ ]

Name: ______________________ Date: ______________________

ID: ______________________ Section: ______________________

Section 16.14: Colligative Properties of Solutions (Optional)

59. Is the partial pressure exerted by one component of a gaseous mixture at a given temperature and volume a colligative property? Justify your answer, pointing out in the process what classified a property as "colligative."

60. 27.2 g of aniline, $C_6H_5NH_2$, is dissolved in 1.20×10^2 g of water. At what temperatures will the solution freeze and boil?

Answer: [ ]

61. Calculate the freezing point of a solution of 2.12 g of naphthalene, $C_{10}H_8$, in 32.0 g of benzene, C_6H_6. Pure benzene freezes at 5.50°C, and its $K_f = 5.10°C/m$.

Answer: [ ]

Name: ______________________ Date: ______________________

ID: ______________________ Section: ______________________

62. What is the molality of a solution of an unknown solute in acetic acid if it freezes at 14.1°C? The normal freezing point of acetic acid is 16.6°C, and K_f = 3.90°C/m.

Answer: [ ]

63. A solution of 16.1 g of an unknown solute in 6.00×10^2 g of water boils at 100.28°C. Find the molar mass of the solute.

Answer: [ ]

64. When 12.4 g of an unknown solute is dissolved in 90.0 g of phenol, the freezing point depression is 9.6°C. Calculate the molar mass of the solute if K_f = 3.56°C/m for phenol.

Answer: [ ]

Name: ______________________ Date: ______________________

ID: ______________________ Section: ______________________

65. The normal freezing point of an unknown solvent is 28.7°C. A solution of 11.4 g of ethanol, C_2H_5OH, in 2.00×10^2 g of the solvent freezes at 22.5°C. What is the molal freezing point constant of the solvent?

Answer: [ ]

General Questions

66. When you heat water on a stove, small bubbles appear long before the water begins to boil. What are they? Explain why they appear.

67. Does percentage concentration of a solution depend on temperature?

More Challenging Problems

68. In Chapter 2 we explain that physical properties must be employed to separate components of a mixture. Suggest a way to separate two immiscible liquids.

Name: ______________________ Date: ______________________

ID: ______________________ Section: ______________________

69. Silver acetate has a solubility of 2.52 g/100 g water at 80°C and 1.02 g/100 g water at 20°C. How do you prepare a supersaturated solution of silver acetate? Why does crystallization not occur as soon as the ion concentration is greater than the concentration of a saturated solution?

70. The density of 18.0% HCl is 1.09 g/mL. Calculate its molarity.

Answer: [ ]

71. Calculate the mass of $H_2C_2O_4 \cdot 2\ H_2O$ required for 2.50×10^2 mL 0.500 N $H_2C_2O_4$ for the reaction $H_2C_2O_4 + 2\ OH^- \rightarrow C_2O_4^{2-} + 2\ H_2O$.

Answer: [ ]

Name: ______________________ Date: ______________________

ID: ______________________ Section: ______________________

72. Sodium carbonate decahydrate is used as a base in the reaction $CO_3^{2-} + 2\ H^+ \rightarrow CO_2 + H_2O$. Calculate the mass of the hydrate needed to prepare 1.00×10^2 mL of 0.500 N sodium carbonate.

Answer: [ ]

73. 25.0 mL of 0.269 M nickel chloride are combined with 30.0 mL 0.260 M potassium hydroxide. How many grams of nickel hydroxide can precipitate?

Answer: [ ]

74. Calculate the hydroxide ion concentration in a 20.00 mL sample of an unknown if 14.75 mL 0.248 M sulfuric acid is used in a neutralization reaction.

Answer: [ ]

Name: ______________________ Date: ______________________

ID: ______________________ Section: ______________________

75. A student received a 599 mg sample of a mixture of sodium hydrogen phosphate and sodium dihydrogen phosphate. She is to find the percentage of each compound in the sample. After dissolving the mixture, she titrated it with 19.58 mL 0.201 M sodium hydroxide. If the only reaction is $NaH_2PO_4 + NaOH \rightarrow Na_2HPO_4 + H_2O$, find the required percentages.

Answer: ______________________

76. A solution has been defined as a homogeneous mixture. Pure air is a solution. Does it follow that the atmosphere is a solution?

77. If you know either the percentage concentration of a solution or its molarity, what additional information must you have before you can convert to the other concentration?

Answers to Target Checks

1. A solution is a homogeneous mixture. Freshly opened root beer is a mixture, but it is not homogeneous because it consists of two states of matter, gas and liquid. Flat root beer satisfies the definition of a solution.

2. a) The solvent is water, the substance present in the greatest amount. The solutes are sugar and red food coloring.

   b) You can make a dilute solution from a concentrated solution by either adding more solvent or by removing some of the solute. You can make a concentrated solution from a dilute solution by either adding more solute or by removing some solvent, perhaps by evaporation.

   c) i) The solution is saturated at 14 g Y in 100 g water.
   ii) The solution is unsaturated from 0 to slightly less than 14 g Y in 100 g water.
   iii) The solution is supersaturated from slightly more than 14 g Y in 100 g water to 30 g Y in 100 g water.

   d) Water and gasoline are immiscible.

3. a) The sugar molecules start locked in fixed positions in a solid crystal. The molecules vibrate in place. When the solid sugar is placed in water, sugar molecules on the edges of the crystal are attracted by water molecules. When a sugar molecule becomes dissolved, the attractions to the water molecules are similar to those that held it in the crystal.

   b) To speed up the dissolving process you could heat the mixture, stir the mixture, and use more finely divided sugar crystals (confectioner's or powdered sugar rather than granulated sugar).

4. a) Water molecules are small and polar, held together by dipole forces and hydrogen bonding. Diesel fuel has large nonpolar molecules held together by dispersion forces. These very different molecules have different intermolecular attractions, so they do not mix.

   b) Bottlers would, in theory, need to increase the amount of CO_2 dissolved in their beverages. The solubility of a gas dissolved in a liquid increases as the partial pressure of that gas above the liquid increases. If its solubility is higher, more gas dissolves before saturation is reached, and it fizzes out.

Chapter 17

Net Ionic Equations

☑ TARGET CHECK 17.1-WB

What property must a solute have to be classified as a strong electrolyte? Weak electrolyte? Nonelectrolyte?

EXAMPLE 17.1-WB

Write the dissolving equation for the following ionic compounds:

Potassium fluoride:

Zinc bromide:

Sodium phosphate:

Write the formula for each solid, followed by a reaction arrow and the formulas of the ions that that will exist in solution when the solid dissolves.

$KF(s) \rightarrow K^+(aq) + F^-(aq)$

$ZnBr_2(s) \rightarrow Zn^{2+}(aq) + 2\ Br^-(aq)$

$Na_3PO_4(s) \rightarrow 3\ Na^+(aq) + PO_4^{3-}(aq)$

EXAMPLE 17.2-WB

Write the formulas of the ions present in solutions of the following ionic compounds: sodium iodide, barium nitrate, lithium sulfate, ammonium phosphate.

Write the dissolving equation, if you wish, but try to just "think" the dissolving equation and write the products.

NaI: $Na^+(aq) + I^-(aq)$ $Ba(NO_3)_2$: $Ba^{2+}(aq) + 2\ NO_3^-(aq)$

Li_2SO_4: $2\ Li^+(aq) + SO_4^{2-}(aq)$ $(NH_4)_3PO_4$: $3\ NH_4^+(aq) + PO_4^{3-}(aq)$

EXAMPLE 17.3-WB

Write the major species in solutions of the following acids: nitric acid, hydrofluoric acid, hydrosulfuric acid, perchloric acid.

Be sure that you've memorized the names and formulas of the seven common strong acids. Complete the example.

HNO_3: $H^+(aq) + NO_3^-(aq)$ HF(aq): HF(aq)

$H_2S(aq)$: $H_2S(aq)$ $HClO_4(aq)$: $H^+(aq) + ClO_4^-(aq)$

EXAMPLE 17.4-WB

A tiny piece of solid potassium is dropped into a solution of hydrochloric acid. A violent reaction ensues, emitting bubbles of hydrogen gas. The solution that remains is aqueous potassium chloride. Write the conventional, total ionic, and net ionic equations for the reaction.

The conventional equation follows from the description of reactants and products in the problem statement. Write each formula, followed by its state symbol, and then balance the equation.

$2\ K(s) + 2\ HCl(aq) \rightarrow H_2(g) + 2\ KCl(aq)$

Next, you need to write the total ionic equation. To do this, replace each aqueous substance that is a strong acid or soluble ionic compound with its major species. There are two aqueous substances in your conventional equation. Classify HCl(aq) and KCl(aq) into the appropriate category, strong acid, weak acid, soluble ionic compound, or insoluble ionic compound, and explain your reasoning in each case.

HCl(aq) is a strong acid, and KCl(aq) is a soluble ionic compound.
HCl(aq) is a strong acid because it is one of the memorized seven common strong acids. KCl(aq) is a soluble ionic compound because the problem statement says that it is a solution.

Write the total ionic equation.

$2\ K(s) + 2\ H^+(aq) + 2\ Cl^-(aq) \rightarrow H_2(g) + 2\ K^+(aq) + 2\ Cl^-(aq)$

This is a good point at which to check for potential errors. In practice, you won't write out this check; you'll do it mentally. Nonetheless, we'll ask you to write out the check just this once in this example to help you learn the net ionic equation writing process. Verify both atom and charge balance.

---

2 K, 2 H, 2 Cl on each side. Atoms balance.
2 positives and 2 negatives on the left; net zero charge. Same on the right. Charge balances.

Now complete the procedure by crossing out the spectator ions. Write the net ionic equation. Check to be sure that it remains balanced.

$2\ K(s) + 2\ H^+(aq) \rightarrow H_2(g) + 2\ K^+(aq)$
2 K and 2 H on each side. Atoms check. 2+ charge on each side. Charge is balanced.
This net ionic equation says that the hydrogen ions from the strong acid react with the solid potassium metal, forming hydrogen gas and potassium ions that are found in solution.

EXAMPLE 17.5-WB

Write the conventional, total ionic, and net ionic equation for the reaction of sodium fluoride solution and hydrochloric acid, yielding sodium chloride solution and hydrofluoric acid.

There are two acids in this reaction, hydrochloric and hydrofluoric. What do you know about how these acids are classified? Are they strong or weak?

Hydrochloric acid is a strong acid, and hydrofluoric acid is weak.

Keep in mind that you do not separate a weak acid into ions, even though it is aqueous, and complete the three steps of the procedure to arrive at the net ionic equation. Also check your net ionic equation for charge and atom balance.

$NaF(aq) + HCl(aq) \rightarrow NaCl(aq) + HF(aq)$
$Na^{+}(aq) + F^{-}(aq) + H^{+}(aq) + Cl^{-}(aq) \rightarrow Na^{+}(aq) + Cl^{-}(aq) + HF(aq)$
$F^{-}(aq) + H^{+}(aq) \rightarrow HF(aq)$
Charges and atoms are balanced.

In this reaction, the hydrogen and fluoride ions combine to make a molecular product, hydrofluoric acid. The sodium and chloride ions remain unchanged in solution.

EXAMPLE 17.6-WB

Solid barium is added to a solution of aluminum nitrate. Write the net ionic equation for the reaction that occurs.

Step 1 is to write the conventional equation. You have to predict the products. Notice that the reactants are a single element and an ionic compound. This means that the potential reaction is a single-replacement reaction. The products will be a single element and an ionic compound, but the reactant element will *replace* an ion in the reactant ionic compound. The ion that gets "kicked out" of the reactant ionic compound becomes the single element product. Write the conventional equation.

$3\ Ba(s) + 2\ Al(NO_3)_3(aq) \rightarrow 3\ Ba(NO_3)_2(aq) + 2\ Al(s)$

The reactant element barium becomes an ion as a product. Since barium is in Group 2A/2, it forms a 2+ ion. Combined with the NO_3^{-} ion, the product ionic compound is $Ba(NO_3)_2$. The reactant aluminum ion changes to elemental aluminum. Once the product formulas are written, the equation must be balanced.

You are ready to write the total ionic equation. Separate the soluble ionic compounds into their ions. Be careful about keeping the correct number of each ion. Mentally check for atom and charge balance after you've written the equation.

$$3\ Ba(s) + 2\ Al^{3+}(aq) + 6\ NO_3^-(aq) \rightarrow 3\ Ba^{2+}(aq) + 6\ NO_3^-(aq) + 2\ Al(s)$$

Both aqueous compounds are separated into their ions. The coefficient of 2 on aluminum nitrate is multiplied by the nitrate ion subscript 3 to result in 6 nitrate ions on the left. A similar 3 × 2 gives 6 nitrate ions on the right.

The final step is to eliminate the spectator ions, which results in the net ionic equation.

$$3\ Ba(s) + 2\ Al^{3+}(aq) \rightarrow 3\ Ba^{2+}(aq) + 2\ Al(s)$$

Six nitrate ions were common to each side of the equation and therefore eliminated to produce the net ionic equation.

EXAMPLE 17.7-WB

Write the three equations for the reaction between nickel and a solution of magnesium chloride, if any.

See what you can do this time without hints.

$$Ni(s) + MgCl_2(aq) \rightarrow NR$$

Nickel is beneath magnesium in the activity series, so no reaction occurs. Remind yourself to get in the habit of checking the activity series every time you have a single-replacement reaction.

EXAMPLE 17.8-WB

If potassium is placed into water, hydrogen gas bubbles out. Write all three equations.

This reaction is more clearly seen if you write the formula of water as HOH. Treat it as a weak acid, with only the first hydrogen ionizable. Go for all three equations, but watch those state symbols.

$2\ K(s) + 2\ HOH(\ell) \rightarrow 2\ KOH(aq) + H_2(g)$

$2\ K(s) + 2\ HOH(\ell) \rightarrow 2\ K^+(aq) + 2\ OH^-(aq) + H_2(g)$

Water is a liquid molecular compound, $HOH(\ell)$. It does not break into ions. Never separate a gas (g), a liquid (ℓ), or a solid (s) into ions. This particular total ionic equation has no spectators, so it is also the net ionic equation.

EXAMPLE 17.9-WB

Solutions of nickel nitrate and sodium hydroxide are combined. A precipitate of nickel hydroxide forms. What is the net ionic equation for the reaction that occurs?

In this example both reactants are ionic compounds. This means that the potential reaction involves "switching partners." Write the conventional equation.

$Ni(NO_3)_2(aq) + 2\ NaOH(aq) \rightarrow Ni(OH)_2(s) + 2\ NaNO_3(aq)$

Notice how the reactant cation-anion plus cation-anion formed product cation-anion plus cation-anion, but the partners were changed.

Write the total ionic and net ionic equations.

$Ni^{2+}(aq) + 2\ NO_3^-(aq) + 2\ Na^+(aq) + 2\ OH^-(aq) \rightarrow Ni(OH)_2(s) + 2\ Na^+(aq) + 2\ NO_3^-(aq)$

$Ni^{2+}(aq) + 2\ OH^-(aq) \rightarrow Ni(OH)_2(s)$

The nickel ion from one solution combines with the hydroxide ion from the other solution to form solid nickel hydroxide.

EXAMPLE 17.10-WB

A sodium sulfite solution is poured into a solution of barium bromide. Barium sulfite precipitates. Write the net ionic equation for the reaction.

Note that two ionic compounds are combined. This indicates a double-replacement reaction. Write the conventional equation.

$Na_2SO_3(aq) + BaBr_2(aq) \rightarrow BaSO_3(s) + 2\ NaBr(aq)$

Write the total ionic and net ionic equations.

$2\ Na^{+}(aq) + SO_3^{2-}(aq) + Ba^{2+}(aq) + 2\ Br^{-}(aq) \rightarrow BaSO_3(s) + 2\ Na^{+}(aq) + 2\ Br^{-}(aq)$
$SO_3^{2-}(aq) + Ba^{2+}(aq) \rightarrow BaSO_3(s)$

EXAMPLE 17.11-WB

Write the net ionic equation for the formation of iron(II) phosphate.

Notice how the problem statement does not give reactants. Iron(II) phosphate is the precipitate, and you are to go directly to the net ionic equation. The reactants are the ions in the product compound.

$3\ Fe^{2+}(aq) + 2\ PO_4^{3-}(aq) \rightarrow Fe_3(PO_4)_2(s)$

EXAMPLE 17.12-WB

Write the net ionic equation for any reaction that will occur when solutions of nickel chloride and ammonium carbonate are combined.

This example does not ask for all three equations, but only the net ionic equation The best way to get there is to write all three equations, however. The precipitate is not identified, so you will need to consult the solubility rules to determine the states of the product compounds.

$NiCl_2(aq) + (NH_4)_2CO_3(aq) \rightarrow NiCO_3(s) + 2\ NH_4Cl(aq)$
If you are using the solubility table, the intersection of the nickel ion and carbonate ion lines indicates that the compound is a solid. The ammonium ion line and the chloride ion line intersect to point out that the compound is aqueous. If you are using the solubility rules, carbonates are usually insoluble, and nickel is not an exception. Ammonium salts are usually soluble; additionally, chlorides are also usually soluble.

Now complete the total and net ionic equations.

$Ni^{2+}(aq) + 2\ Cl^{-}(aq) + 2\ NH_4^{+}(aq) + CO_3^{2-}(aq) \rightarrow NiCO_3(s) + 2\ NH_4^{+}(aq) + 2\ Cl^{-}(aq)$
$Ni^{2+}(aq) + CO_3^{2-}(aq) \rightarrow NiCO_3(s)$

EXAMPLE 17.13-WB

Write the net ionic equation for any reaction that occurs when solutions of ammonium sulfate and potassium nitrate are combined.

Again, be careful.

$(NH_4)_2SO_4(aq) + KNO_3(aq) \rightarrow NR$

This time both new combinations of ions are soluble, so there is no precipitation reaction. If you complete the conventional equation, you will find that *all* ions are spectators.

EXAMPLE 17.14-WB

Barium hydroxide and nitric acid solutions are combined. Write the conventional, total ionic, and net ionic equations.

Start with the conventional equation.

$Ba(OH)_2(aq) + 2\ HNO_3(aq) \rightarrow Ba(NO_3)_2(aq) + 2\ H_2O(\ell)$

Write the total ionic and net ionic equations. Be careful with the final net ionic equation.

$Ba^{2+}(aq) + 2\ OH^-(aq) + 2\ H^+(aq) + 2\ NO_3^-(aq) \rightarrow Ba^{2+}(aq) + 2\ NO_3^-(aq) + 2\ H_2O(\ell)$

$2\ OH^-(aq) + 2\ H^+(aq) \rightarrow 2\ H_2O(\ell)$

$OH^-(aq) + H^+(aq) \rightarrow H_2O(\ell)$

The final equation must be reduced to the lowest coefficients.

EXAMPLE 17.15-WB

Write the conventional, total ionic, and net ionic equations for the reaction that occurs when a solution of nitrous acid is combined with a potassium hydroxide solution.

Start with the conventional equation.

$HNO_2(aq) + KOH(aq) \rightarrow H_2O(\ell) + KNO_2(aq)$

Keeping in mind that nitrous acid is a weak acid, write the total ionic and net ionic equations.

$HNO_2(aq) + K^+(aq) + OH^-(aq) \rightarrow H_2O(\ell) + K^+(aq) + NO_2^-(aq)$

$HNO_2(aq) + OH^-(aq) \rightarrow H_2O(\ell) + NO_2^-(aq)$

EXAMPLE 17.16-WB

Solutions of potassium acetate (acetate ion, $C_2H_3O_2^-$) and chloric acid are mixed. Develop the net ionic equation for the reaction that occurs.

Write all three equations.

$KC_2H_3O_2(aq) + HClO_3(aq) \rightarrow KClO_3(aq) + HC_2H_3O_2(aq)$
$K^+(aq) + C_2H_3O_2^-(aq) + H^+(aq) + ClO_3^-(aq) \rightarrow K^+(aq) + ClO_3^-(aq) + HC_2H_3O_2(aq)$
$C_2H_3O_2^-(aq) + H^+(aq) \rightarrow HC_2H_3O_2(aq)$

EXAMPLE 17.17-WB

Write the net ionic equation for combinations of the following solutions:
(a) Perchloric acid and sodium hydroxide
(b) Sodium fluoride and hydroiodic acid

Write the formulas of all four ions found in the two solutions of Part (a).

$H^+(aq)$, $ClO_4^-(aq)$, $Na^+(aq)$, and $OH^-(aq)$

Now consider the potential combinations of cations and anions. What molecular product will form? Write the net ionic equation for the formation of that product.

$H^+(aq) + OH^-(aq) \rightarrow H_2O(\ell)$

Complete Part (b).

$F^-(aq) + H^+(aq) \rightarrow HF(aq)$
Hydrofluoric acid is weak, and thus it forms when hydrogen ions and fluoride ions are combined.

EXAMPLE 17.18-WB

Solutions of magnesium sulfite and hydrochloric acid are mixed. Write the conventional, total ionic, and net ionic equations for the reaction.

Go for all three equations. Watch for unstable products.

$MgSO_3(aq) + 2\ HCl(aq) \rightarrow MgCl_2(aq) + \text{“}H_2SO_3(aq)\text{”}$
$MgSO_3(aq) + 2\ HCl(aq) \rightarrow MgCl_2(aq) + SO_2(aq) + H_2O(\ell)$
$Mg^{2+}(aq) + SO_3^{2-}(aq) + 2\ H^+(aq) + 2\ Cl^-(aq) \rightarrow Mg^{2+}(aq) + 2\ Cl^-(aq) + SO_2(aq) + H_2O(\ell)$
$SO_3^{2-}(aq) + 2\ H^+(aq) \rightarrow SO_2(aq) + H_2O(\ell)$

The predicted but unstable $H_2SO_3(aq)$ in the first step is replaced with $SO_2(aq) + H_2O(\ell)$ in the second step. This is the correct conventional equation. The total ionic and net ionic equations follow in the usual manner.

EXAMPLE 17.19-WB

Potassium hydrogen phthalate, $KHC_8H_4O_4$, is an ionic compound with an anion that has an acidic hydrogen. It is used in analytical laboratories to determine the concentration of sodium hydroxide solutions. Write the net ionic equation for the reaction of solid potassium hydrogen phthalate and a solution of sodium hydroxide.

Write the conventional equation. It is permissible to have two different cations balance the negative charge of a single anion.

$KHC_8H_4O_4(s) + NaOH(aq) \rightarrow NaKC_8H_4O_4(aq) + H_2O(\ell)$

You may have written $KNaC_8H_4O_4(aq)$, which is acceptable at this point in your studies. The important concept is to have two 1+ ions to balance the charge of the 2– phthalate ion. The double replacement cation exchange is between the acidic hydrogen and the sodium ion.

Now write the total ionic equation. Carefully follow the rules for breaking soluble ionic compounds into their ions when writing the equation.

$KHC_8H_4O_4(s) + Na^+(aq) + OH^-(aq) \rightarrow Na^+(aq) + K^+(aq) + C_8H_4O_4^{2-}(aq) + H_2O(\ell)$

The net ionic equation completes the problem.

$KHC_8H_4O_4(s) + OH^-(aq) \rightarrow K^+(aq) + C_8H_4O_4^{2-}(aq) + H_2O(\ell)$

When solid $KHC_8H_4O_4$ is placed into an aqueous solution, it dissolves: $KHC_8H_4O_4(s) \rightarrow K^+(aq) + HC_8H_4O_4^-(aq)$. The acid anion then reacts with hydroxide ion: $HC_8H_4O_4^-(aq) + OH^-(aq) \rightarrow C_8H_4O_4^{2-}(aq) + H_2O(\ell)$. The net ionic equation summarizes both steps.

EXAMPLE 17.20-WB

Magnesium chloride and iron(III) nitrate solutions are combined. Write the net ionic equation for the reaction that occurs, if any.

Take it all the way.

There is no reaction.

$3\ MgCl_2(aq) + 2\ Fe(NO_3)_3(aq) \rightarrow 3\ Mg(NO_3)_2(aq) + 2\ FeCl_3(aq)$

After writing the conventional equation, you may have realized that all reactants and products are soluble ionic compounds, and therefore there would be no chemical change. If you took it to the next step, $3\ Mg^{2+}(aq) + 6\ Cl^-(aq) + 2\ Fe^{3+}(aq) + 6\ NO_3^-(aq) \rightarrow 3\ Mg^{2+}(aq) + 6\ NO_3^-(aq) + 2\ Fe^{3+}(aq) + 6\ Cl^-(aq)$, you would see that both sides of the equation are the same.

Name: ______________________ Date: ______________________

ID: ______________________ Section: ______________________

Questions, Exercises, and Problems

Section 17.1: Electrolytes and Solution Conductivity

1. How does a weak electrolyte differ from a nonelectrolyte?

2. How can it be that all soluble ionic compounds are electrolytes but soluble molecular compounds may or may not be electrolytes?

Sections 17.2 *and* 17.3: Solutions of Ionic Compounds *and* Strong and Weak Acids

Questions 3 through 6: Write the major species in the water solution of each substance given. All ionic compounds given are soluble.

3. $(NH_4)_2SO_4$, $MnCl_2$

4. $NiSO_4$, K_3PO_4

5. HNO_3, HBr

Name: ______________________ Date: ______________________

ID: ______________________ Section: ______________________

6. $H_2C_4H_4O_4$, HF

Section 17.4: Net Ionic Equations: What They Are and How to Write Them

Questions 7 to 9: For each reaction described, write the net ionic equation.

7. A zinc chloride solution is mixed with a sodium phosphate solution, forming a precipitate of solid zinc phosphate and a sodium chloride solution.

8. Solid iron metal is dropped into a solution of hydrochloric acid. Hydrogen gas bubbles out, leaving a solution of iron(III) chloride.

9. Aqueous solutions of oxalic acid and sodium chloride form when solid sodium oxalate (oxalate ion, $C_2O_4^{2-}$) is sprinkled into hydrochloric acid.

Name: ______________________ Date: ______________________

ID: ______________________ Section: ______________________

Section 17.5: Single-Replacement Oxidation–Reduction (Redox) Reactions

Questions 10 through 12: For each pair of reactants, write the net ionic equation for any single-replacement redox reaction that may be predicted by Table 17.2 (Section 17.5). If no redox reaction occurs, write NR.

10. $Cu(s) + Li_2SO_4(aq)$

11. $Ba(s) + HCl(aq)$

12. $Ni(s) + CaCl_2(aq)$

Name: ______________________ Date: ______________________

ID: ______________________ Section: ______________________

Section 17.6: Ion Combinations That Form Precipitates

Questions 13 through 16: For each pair of reactants given, write the net ionic equation for any precipitation reaction that may be predicted by Tables 17.3 and 17.4 (Section 17.6). If no precipitation reaction occurs, write NR.

13. $Pb(NO_3)_2(aq) + KI(aq)$

14. $KClO_3(aq) + Mg(NO_2)_2(aq)$

15. $AgNO_3(aq) + LiBr(aq)$

16. $ZnCl_2(aq) + Na_2SO_3(aq)$

Name: ______________________ Date: ______________________

ID: ______________________ Section: ______________________

17. Write the net ionic equations for the precipitation of each of the following insoluble ionic compounds from aqueous solutions: $PbCO_3$; $Ca(OH)_2$.

Section 17.7: Ion Combinations That Form Molecules

Questions 18 through 20: For each pair of reactants given, write the net ionic equation for the molecule-formation reaction that will occur.

18. $NaNO_2(aq)$, $HI(aq)$

19. $KC_3H_5O_3(aq)$, $HClO_4(aq)$

20. $RbOH(aq)$, $HF(aq)$

Name: ______________________ Date: ______________________

ID: ______________________ Section: ______________________

Section 17.8: Ion Combinations That Form Unstable Products

Questions 21 and 22: For each pair of reactants given, write the net ionic equation for the reaction that will occur.

21. $(NH_4)_2SO_3(aq)$, $HBr(aq)$

22. $Na_2SO_3(aq)$, $HClO_3(aq)$

The remaining questions include all types of reactions discussed in this chapter. Use the activity series and solubility rules to predict whether redox or precipitation reactions will take place. If it will, write the net ionic equation; if not, write NR.

23. Barium chloride and sodium sulfite solutions are combined in an oxygen-free atmosphere.

Name: ____________________ Date: ____________________

ID: ____________________ Section: ____________________

24. Copper(II) sulfate and sodium hydroxide solutions are combined.

25. Bubbles appear as hydrochloric acid is poured onto solid magnesium carbonate.

26. Nitric acid appears to "dissolve" solid lead(II) hydroxide.

27. Benzoic acid, $HC_7H_5O_2(s)$, is neutralized by sodium hydroxide solution.

Name: ______________________ Date: ______________________

ID: ______________________ Section: ______________________

28. Nickel is placed into hydrochloric acid.

29. Hydrochloric acid is poured into a solution of sodium hydrogen sulfite.

30. Solutions of magnesium sulfate and ammonium bromide are combined.

31. Magnesium ribbon is placed in hydrochloric acid.

Name: ______________________ Date: ______________________

ID: ______________________ Section: ______________________

32. Solid nickel hydroxide is apparently readily "dissolved" by hydrobromic acid.

33. Sodium fluoride solution is poured into nitric acid.

34. Silver wire is dropped into hydrochloric acid.

35. When metallic lithium is added to water, hydrogen is released.

Name: ______________________ Date: ______________________

ID: ______________________ Section: ______________________

36. Aluminum shavings are dropped into a solution of copper(II) nitrate.

Answers to Target Checks

1. To be classified as a strong electrolyte, a solute must dissociate into ions; its solution will contain ions. A weak electrolyte has a few ions in solution, but it mainly consists of un-ionized molecules. A nonelectrolyte yields a solution having only neutral molecules.

Chapter 18

Acid–Base (Proton-Transfer) Reactions

☑ TARGET CHECK 18.1-WB

Solution A contains more hydrogen ions than hydroxide ions. Solution B contains more hydroxide ions than hydrogen ion. When the two solutions are combined, the resulting solution contains equal numbers of hydrogen and hydroxide ions. (i) Which solution is an Arrhenius acid? (ii) Which solution is an Arrhenius base? (iii) Write the net ionic equation for the reaction that occurs when the solutions are combined.

☑ TARGET CHECK 18.2-WB

Identify the acid, base, proton receiver, and proton donor in the following reaction: $HBr(g) + H_2O(\ell) \rightarrow H_3O^+(aq) + Cl^-(aq)$.

☑ TARGET CHECK 18.3-WB

Write the net ionic equation for the reaction between an ammonia solution and perchloric acid.

☑ TARGET CHECK 18.4-WB

(a) From the following, state which can be Lewis acids and which can be Lewis bases: Ca^{2+}, F^-, Na^+, CHO_2^-, Fe^{3+}, NH_3.

(b) Consider the reaction between sliver ion and ammonia:

$$Ag^+(aq) + 2\ NH_3(aq) \rightleftharpoons [H_3N{-}Ag{-}NH_3]^+(aq)$$

Define the terms *Lewis acid* and *Lewis base*. Identify each in this reaction.

EXAMPLE 18.1-WB

(a) Write the formula of the conjugate base of each of the following acids: H_2SO_4, HCO_3^-, H_2O.
(b) Write the formula of the conjugate acid of each of the following bases: PO_4^{3-}, HCO_3^-, H_2O.

To write the formula of a conjugate base of an acid, remove an H^+ ion. Complete Part (a).

(a) H_2SO_4: HSO_4^-; HCO_3^-: CO_3^{2-}; H_2O: OH^-

To write the formula of a conjugate acid of a base, add an H^+ ion. Complete Part (b).

(b) PO_4^{3-}: HPO_4^{2-}; HCO_3^-: H_2CO_3; H_2O: H_3O^+

EXAMPLE 18.2-WB

Identify conjugate acid-base pairs in the equation $HC_2H_3O_2(aq) + HPO_4^{2-}(aq) \rightleftharpoons C_2H_3O_2^-(aq) + H_2PO_4^-(aq)$. Specify the acid and the base in each pair.

$HC_2H_3O_2(aq)$ [acid] and $C_2H_3O_2^-(aq)$ [base]
$HPO_4^{2-}(aq)$ [base] and $H_2PO_4^-(aq)$ [acid]

EXAMPLE 18.3-WB

List the following acids in order of increasing strength: chloric acid, acetic acid, water.

You may refer to Table 18.1 on Page 509 of the textbook.

$H_2O < HC_2H_3O_2 < HClO_3$

EXAMPLE 18.4-WB

List the following bases in order of decreasing strength: hydrogen sulfide ion, hydrogen oxalate ion, sulfite ion.

Find the necessary additional information in Table 18.1 on Page 509 of the textbook.

$SO_3^{2-} > HS^- > HC_2O_4^-$

EXAMPLE 18.5-WB

Write the net ionic equation for the acid-base reaction between hydrogen carbonate ion and hypochlorite ion and predict which side will be favored at equilibrium.

The first step is to write and balance a single-proton transfer reaction.

$HCO_3^-(aq) + ClO^-(aq) \rightleftharpoons CO_3^{2-}(aq) + HClO(aq)$

Now label the acid and base on each side of the equation. Use A for acid and B for base. Write the letters under your equation in the top frame.

$HCO_3^-(aq) + ClO^-(aq) \rightleftharpoons CO_3^{2-}(aq) + HClO(aq)$
A B B A

Modify your acid and base labels to indicate the stronger acids and bases. Write S before each A or B to indicate the stronger acid and W before the As and Bs to label the weaker acids and bases. Use Table 18.1.

$HCO_3^-(aq) + ClO^-(aq) \rightleftharpoons CO_3^{2-}(aq) + HClO(aq)$
WA WB SB SA

Which side of the equation will be favored at equilibrium?

The reverse reaction will be favored.
This conclusion is based on HCO_3^- being a weaker acid than HClO and ClO^- being a weaker base than CO_3^{2-}.

EXAMPLE 18.6-WB

What is the hydrogen ion concentration in a solution in which $[OH^-] = 10^{-4}$ M? Is the solution acidic or basic?

GIVEN: $[OH^-] = 10^{-4}$ M *WANTED:* $[H^+]$

EQUATION: $[H^+] = \frac{K_w}{[OH^-]} = \frac{10^{-14}}{10^{-4}} = 10^{-10}$ M

Since $[OH^-] = 10^{-4}$ M > 10^{-10} M = $[H^+]$, the solution is basic.

EXAMPLE 18.7-WB

Complete the blanks in each of the following.

pH = 8 $[H^+]$ = ____________

pOH = 11 $[OH^-]$ = ____________

$[H^+] = 10^{-5}$ pH = ____________

$[OH^-] = 10^{-2}$ pOH = ____________

---

pH = 8 $[H^+] = 10^{-8}$ M

pOH = 11 $[OH^-] = 10^{-11}$ M

$[H^+] = 10^{-5}$ pH = 6

$[OH^-] = 10^{-2}$ pOH = 2

EXAMPLE 18.9-WB

A solution has pH = 4. Calculate the $[H^+]$, $[OH^-]$, and pOH of this solution.

$$\text{pH} = 4;\ [H^+] = 10^{-4}\ \text{M};\ [OH^-] = \frac{10^{-14}}{10^{-4}} = 10^{-10};\ \text{pOH} = 10$$

EXAMPLE 18.10-WB

Consider the following solutions and their respective pH values:

Solution	pH value	Solution	pH value
A	1	C	4
B	9	D	11

Arrange these solutions in order of increasing basicity (least basic first).

There are a number of correct approaches to this problem. Since all values are already listed as pH, we suggest that you think in terms of pH to solve the problem. Complete the blank in the following statement: the higher the pH value, the ___ (more or less) basic the solution.

The higher the pH value, the more basic the solution.

Now list the letters of the solutions from least basic to most basic.

A < C < B < D

EXAMPLE 18.12-WB

The hydrogen ion concentration of a solution is 6.2×10^{-5}. Calculate the pH.

$$\text{pH} = -\log [H^+] = -\log (6.2 \times 10^{-5}) = 4.21$$

EXAMPLE 18.14-WB

What is the hydroxide ion concentration of a solution with pH = 8.55?

You cannot go directly from pH to $[OH^-]$. You can either convert to $[H^+]$ and then to $[OH^-]$ or you can go to pOH and then to $[OH^-]$. We'll show both approaches in our solution.

$[OH^-] = 3.5 \times 10^{-6}$ M or 3.6×10^{-6} M

$[H^+]$ = antilog (–8.55) = 2.8×10^{-9} $[OH^-] = \frac{K_w}{[H^+]} = \frac{1.0 \times 10^{-14}}{2.8 \times 10^{-9}} = 3.6 \times 10^{-6}$ M *or*

pOH = 14.00 – pH = 14.00 – 8.55 = 5.45 $[OH^-]$ = antilog (–5.45) = 3.5×10^{-6} M

EXAMPLE 18.15-WB

The pOH of a solution is 3.47. Calculate the pH, $[H^+]$, and $[OH^-]$ of this solution.

You can work through the pH loop in either direction.

---

pH = 10.53; $[OH^-] = 3.4 \times 10^{-4}$ M; $[H^+] = 3.0 \times 10^{-11}$ M

pH = 14.00 – pOH = 14.00 – 3.47 = 10.53

$[OH^-]$ = antilog (–pOH) = antilog (–3.47) = 3.4×10^{-4} M

$[H^+]$ = antilog (–pH) = antilog (–10.53) = 3.0×10^{-11} M

Name: ______________________ Date: ______________________

ID: ______________________ Section: ______________________

Questions, Exercises, and Problems

Section 18.1: The Arrhenius Theory of Acids and Bases
Section 18.2: The Brønsted–Lowry Theory of Acids and Bases
Section 18.3: The Lewis Theory of Acids and Bases (Optional)

1. Identify at least two of the classical properties of acids and two of bases. For one acid property and one base property, show how it is related to the ion associated with an acid or base.

2. Distinguish between an Arrhenius base and a Brønsted–Lowry base. Are the two concepts in agreement? Justify your answer.

3. Explain or illustrate by an example what is meant by identifying a Lewis acid as an electron-pair acceptor and a Lewis base as an electron-pair donor.

Name: ______________________ Date: ______________________

ID: ______________________ Section: ______________________

4. Diethyl ether reacts with boron trifluoride by forming a covalent bond between the molecules. Describe the reaction from the standpoint of the Lewis acid–base theory, based on the following "structural" equation:

$$BF_3 + :O(C_2H_5)_2 \longrightarrow F_3B{-}O(C_2H_5)_2$$

Section 18.4: Conjugate Acid–Base Pairs

5. Give the formula of the conjugate base of HF; of $H_2PO_4^-$. Give the formula of the conjugate acid of NO_2^-; of $H_2PO_4^-$.

6. For the reaction $HSO_4^-(aq) + C_2O_4^{2-}(aq) \rightleftharpoons SO_4^{2-}(aq) + HC_2O_4^-(aq)$ identify the acid and the base on each side of the equation—that is, the acid and base for the forward reaction and the acid and base for the reverse reaction.

Name: ______________________ Date: ______________________

ID: ______________________ Section: ______________________

7. Identify the conjugate acid–base pairs in Question 6.

8. For the reaction $HNO_2(aq) + C_3H_5O_2^-(aq) \rightleftharpoons NO_2^-(aq) + HC_3H_5O_2(aq)$ identify both conjugate acid–base pairs.

9. Identify the conjugate acid–base pairs in $NH_4^+(aq) + HPO_4^{2-}(aq) \rightleftharpoons NH_3(aq) + H_2PO_4^-(aq)$.

Section 18.5: Relative Strengths of Acids and Bases

Refer to Table 18.1 (textbook Page 509) when answering questions in this section.

10. What is the difference between a strong base and a weak base, according to the Brønsted-Lowry concept? Identify two examples of strong bases and two examples of weak bases.

11. List the following acids in order of their increasing strength (weakest acid first): $HC_2O_4^-$, H_2SO_3, H_2O, $HClO$.

Name: ______________________ Date: ______________________

ID: ______________________ Section: ______________________

12. List the following bases in order of their decreasing strength (strongest base first): CN^-, H_2O, HCO_3^-, ClO^-, Cl^-.

Section 18.6: Predicting Acid–Base Reactions

For each acid and base given in this section, complete a proton-transfer equation for the transfer of one proton. Using Table 18.1, predict the direction in which the resulting equilibrium will be favored.

13. $HC_3H_5O_2(aq) + PO_4^{3-}(aq) \rightleftharpoons$

14. $HSO_4^-(aq) + CO_3^{2-}(aq) \rightleftharpoons$

15. $H_2CO_3(aq) + NO_3^-(aq) \rightleftharpoons$

16. $NO_2^-(aq) + H_3O^+(aq) \rightleftharpoons$

17. $HSO_4^-(aq) + HC_2O_4^-(aq) \rightleftharpoons$

Section 18.8: The Water Equilibrium

18. Of what significance is the very small value of 10^{-14} for K_w, the ionization equilibrium constant for water?

19. $[H^+] = 10^{-5}$ M and $[OH^-] = 10^{-9}$ M in a certain solution. Is the solution acidic, basic, or neutral? How do you know?

Name: ______________________ Date: ______________________

ID: ______________________ Section: ______________________

20. What is $[OH^-]$ in 0.01 M HCl? (*Hint:* Begin by finding $[H^+]$ in 0.01 M HCl.)

Answer:

Section 18.9: pH and pOH (Integer Values Only)

21. In which classification identified in textbook Question 21, strongly acidic, weakly acidic, strongly basic, weakly basic, and neutral, or close to neutral, does each of the following solutions belong: (a) pH = 7, (b) pH = 9, (c) pOH = 3?

22. If the pH of a solution is 8.6, is the solution acidic or basic? How do you reach your conclusion? List in order the pH values of a solution that is neutral, one that is basic, and one that is acidic.

Questions 23 to 26: The pH, pOH, $[H^+]$, or $[OH^-]$ is given. Find each of the other values. Classify each solution as strongly acidic, weakly acidic, neutral (or close to neutral), weakly basic, or strongly basic.

23. pH = 5

Answer:

Name: ____________________ Date: ____________________

ID: ____________________ Section: ____________________

24. $[OH^-] = 10^{-1}$ M

Answer:

25. pOH = 4

Answer:

26. $[H^+] = 10^{-9}$ M

Answer:

Name: ______________________________ Date: ______________________

ID: ________________________________ Section: ____________________

Section 18.10: Noninteger pH–$[H^+]$ and pOH–$[OH^-]$ Conversions (Optional)

Questions 27 to 30: The pH, pOH, $[H^+]$, or $[OH^-]$ of a solution is given. Find each of the other values.

27. $[OH^-] = 2.5 \times 10^{-10}$ M

Answer:

28. pH = 4.06

Answer:

29. $[H^+] = 2.8 \times 10^{-1}$ M

Answer:

Name: ______________________ Date: ______________________

ID: ______________________ Section: ______________________

30. pOH = 7.40

Answer:

General Questions

31. Can a substance that does not have hydrogen atoms be a Brønsted-Lowry acid? Explain.

32. Why do you suppose chemists prefer the pH scale to expressing hydrogen ion concentration directly? In other words, what is the advantage of saying pH = 4 rather than $[H^+] = 10^{-4}$?

33. What is the bromide ion concentration in a solution if a report gives the pBr as 7.2?

Name: ______________________ Date: ______________________

ID: ______________________ Section: ______________________

More Challenging Questions

34. Nonmetal oxides can react with water to form acids. For example, carbon dioxide reacts with water to form carbonic acid: $CO_2 + H_2O \rightarrow H_2CO_3$. What acid forms as the result of the reaction of sulfur trioxide and water? Write the equation for the reaction.

35. Metal oxides can react with water to form bases. Write an equation to show how calcium oxide can react with water to form a base. Name the base.

36. Nitrogen dioxide is emitted in automobile exhaust. Explain how this can contribute to acid rain, which is rain with a low pH.

Answers to Target Checks

1. (i) Solution A is an acid because it contains more hydrogen ions than hydroxide ions.
(ii) Solution B is a base because it contains more hydroxide ions than hydrogen ions.
(iii) $H^+ + OH^- \rightarrow H_2O$

2. Acid and proton donor: HBr(g); base and proton receiver: $H_2O(\ell)$

3. $NH_3(aq) + HClO_4(aq) \rightleftharpoons NH_4^+(aq) + ClO_4^-(aq)$

4. (a) Lewis acids: Ca^{2+}, Na^+, Fe^{3+}; Lewis bases: F^-, CHO_2^-, NH_3
(b) A Lewis acid is an electron-pair acceptor, Ag^+. A Lewis base is an electron-pair donor, NH_3.

Chapter 19

Oxidation–Reduction (Redox) Reactions

☑ TARGET CHECK 19.1-WB

Under what conditions is a rechargeable nickel-cadmium cell a voltaic cell? Under what conditions is it an electrolytic cell?

EXAMPLE 19.1-WB

Combine the following two descriptions of half-reactions to produce a balance redox equation. Identify the oxidation half-reaction and the reduction half-reaction.

Solid copper forms an aqueous solution of copper(I) ion and an electron
Gold(I) ion in aqueous solution gains an electron to form solid gold (gold, Z = 79)

First, you need to translate the written descriptions into formulas and half-reaction equations. Then label the oxidation half-reaction and the reduction half-reaction. Finally, add the two half reactions to get the overall redox reaction net ionic equation.

Oxidation:	$Cu(s)$	$\rightarrow Cu^{+}(aq) + e^{-}$
Reduction:	$Au^{+}(aq) + e^{-}$	$\rightarrow Au(s)$
Redox:	$Cu(s) + Au^{+}(aq)$	$\rightarrow Cu^{+}(aq) + Au(s)$

EXAMPLE 19.2-WB

Classify each of the following equations as an oxidation half-reaction equation or a reduction half-reaction equation:

(a) $ClO_2(aq) + e^- \rightarrow ClO_2^-$
(b) $2\ I^-(aq) \rightarrow I_2(aq) + 2\ e^-$

Combine the half-reaction equations to give a balanced redox reaction.

Consider the classification part of the question first. Classify each half-reaction as either oxidation or reduction, and briefly explain your reasoning.

(a) Reduction because the reactant gains an electron.
(b) Oxidation because the reactant loses an electron.

Now add the half-reactions to get the net ionic equation for the overall reaction. Don't forget to balance the electrons transferred.

$$2\ [ClO_2(aq) + e^- \rightarrow ClO_2^-]$$
$$2\ I^-(aq) \rightarrow I_2(aq) + 2\ e^-$$
$$2\ ClO_2(aq) + 2\ I^-(aq) \rightarrow ClO_2^- + I_2(aq)$$

EXAMPLE 19.3-WB

Combine the following two half-reaction equations to produce a balanced redox equation. Identify the oxidation half-reaction and the reduction half-reaction.

$Au^{3+}(aq) + 3\ e^- \rightarrow Au(s)$ (Au, gold, Z = 79)
$Sn(s) \rightarrow Sn^{2+}(aq) + 2\ e^-$

Reduction: $2\ [Au^{3+}(aq) + 3\ e^- \rightarrow Au(s)]$
Oxidation: $3\ [Sn(s) \rightarrow Sn^{2+}(aq) + 2\ e^-]$
Redox: $2\ Au^{3+}(aq) + 3\ Sn(s) \rightarrow 2\ Au(s) + 3\ Sn^{2+}(aq)$

The gold(III) ion reduction was multiplied by 2, and the tin oxidation was multiplied by 3 to transfer six electrons total for each half-reaction.

EXAMPLE 19.4-WB

Give the oxidation number of each element in each formula: (a) Ca^{2+}, (b) AsO_2^-.

Review the oxidation number rules in the summary box on Page 533, if necessary, before answering this question. Part (a) is governed by Rule 2. Part (b) requires the application of Rules 3 and 5.

(a) +2, (b) As: +3, O: –2

In part (b), the oxidation number of oxygen is –2 (Rule 3). There are two oxygen atoms, for a total of –4. The oxidation numbers must sum to the charge on the ion (Rule 5), –1. The oxidation number of As is therefore +3: $+3 + 2(-2) = -1$.

EXAMPLE 19.5-WB

Give the oxidation number of each element in the formula of each compound: (a) aluminum sulfide, (b) ammonium bromide.

With ionic compounds such as these, consider each ion separately.

(a) Al: +3, S: –2; (b) N: –3, H: +1, Br: –1

Aluminum sulfide is composed of Al^{3+} ions and S^{2-} ions. The oxidation number of an ion is the same as the charge on that ion. Ammonium bromide is made up of NH_4^+ and Br^- ions. In NH_4^+, each hydrogen atom is +1, and the sum of the oxidation numbers is +1. Nitrogen therefore has an oxidation number of –3: $x + 4 = 1$; $x = -3$.

EXAMPLE 19.6-WB

In each reaction, identify the element oxidized or reduced and note the change in oxidation number. For example, for $Pb^{2+}(aq) + 2\ e^- \rightarrow Pb(s)$, you would say that "lead is reduced from +2 to 0."

(a) $K(s) \rightarrow K^+(aq) + e^-$

(b) $2\ H^+(aq) + ClO_4^-(aq) \rightarrow ClO_3^-(aq) + H_2O(\ell)$

In part (b), you'll probably find it helpful to write the oxidation numbers of all the elements so that you can find the pair that changes.

(a) Potassium is oxidized from 0 to +1.

(b) Chlorine is reduced from +7 to +5.

☑ TARGET CHECK 19.2-WB

Explain how oxidation and reduction are defined in terms of oxidation numbers.

EXAMPLE 19.7-WB

In the reaction $Cu^{2+}(aq) + Mg(s) \rightarrow Cu(s) + Mg^{2+}(aq)$, identify: (a) the substance oxidized, (b) the oxidizing agent, (c) the substance reduced, and (d) the reducing agent. In each case, explain how you identified the substance.

(a) Magnesium is oxidized, increasing in oxidation number from 0 to 2.
(b) Copper(II) ion is the oxidizing agent, removing electrons from Mg.
(c) Copper(II) ion is reduced, decreasing in oxidation number from +2 to 0.
(d) Magnesium is the reducing agent, supplying electrons to copper(II) ion.

☑ TARGET CHECK 19.3-WB

Match the letter of the choice below that best describes each of the following:

i) strong attraction for electrons	a	b	c	d	(circle one)
ii) weak attraction for electrons	a	b	c	d	
iii) gives up electrons weakly	a	b	c	d	
iv) gives up electrons freely	a	b	c	d	

a = strong oxidizing agent; b = weak oxidizing agent; c = strong reducing agent; d = weak reducing agent

EXAMPLE 19.8-WB

Which is the stronger oxidizing agent, $Br_2(\ell)$ or $Zn^{2+}(aq)$? Which is the stronger reducing agent, $Br^-(aq)$ or Zn(s)? Write the net ionic equation for the redox reaction between liquid bromine and solid zinc.

Use Table 19.2, Relative Strengths of Oxidizing and Reducing Agents, which can be found on textbook Page 537, to answer the first and second parts.

$Br_2(\ell)$ is a stronger oxidizing agent than $Zn^{2+}(aq)$. Zn(s) is a stronger reducing agent than $Br^-(aq)$.

Now write the oxidation and reduction half-reactions. All reactions in Table 19.2 are written as reduction half-reactions, so you will have to reverse the zinc/zinc ion half-reaction.

Oxidation: $Zn(s) \rightleftharpoons Zn^{2+}(aq) + 2\ e^-$

Reduction: $Br_2(\ell) + 2\ e^- \rightleftharpoons 2\ Br^-(aq)$

Combine the half-reactions to get the net ionic equation for the overall reaction. Label the stronger and weaker oxidizing agents and reducing agents.

$Zn(s) + Br_2(\ell) \rightleftharpoons Zn^{2+}(aq) + 2\ Br^-(aq)$

SRA SOA WOA WRA

The net ionic equation results from summing the two half-reactions. Note how the stronger reducing and oxidizing agents are on the same side of the equilibrium and the weaker oxidizing and reducing agents are together on the other side of the equilibrium. We'll discuss the significance of this when you return to the textbook.

EXAMPLE 19.9-WB

Write the redox reaction between metallic aluminum and a strong acid, $H^+(aq)$, to form aluminum ions and hydrogen gas. Indicate the direction in which the reaction is favored.

2 × Oxidation: $2\,[Al(s) \rightleftharpoons Al^{3+}(aq) + 3\,e^-]$

3 × Reduction: $3\,[2\,H^+(aq) + 2\,e^- \rightleftharpoons H_2(g)]$

Redox: $2\,Al(s) + 6\,H^+(aq) \rightleftharpoons 2\,Al^{3+}(aq) + 3\,H_2(g)$

SRA SOA WOA WRA

The forward direction is favored.

EXAMPLE 19.10-WB

For the reactants $Cl_2(g)$ and Na(s), select from Table 19.2 the needed half-reaction equations, write them so they may be added to produce the equation for the redox reaction, and complete the addition. Predict the direction in which the reaction is favored.

Reduction: $Cl_2(g) + 2\,e^- \rightleftharpoons 2\,Cl^-(aq)$

2 × Oxidation: $2\,[Na(s) \rightleftharpoons Na^+(aq) + e^-]$

Redox: $Cl_2(g) + 2\,Na(s) \rightleftharpoons 2\,Cl^-(aq) + 2\,Na^+(aq)$

SOA SRA WRA WOA

The forward direction is favored.

EXAMPLE 19.11-WB

Consider the reaction of nitrate ion and chloride ion in acidic solution. Write the half-reaction equations and sum them to get the net ionic equation for the overall reaction. Predict the direction in which the reaction is favored.

Reduction: $NO_3^-(aq) + 4\ H^+(aq) + 4\ e^- \rightleftharpoons NO(g) + 2\ H_2O(\ell)$

$2 \times$ Oxidation: $2\ [2\ Cl^-(aq) \rightleftharpoons Cl_2(g) + 2\ e^-]$

Redox: $NO_3^-(aq) + 4\ H^+(aq) + 4\ Cl^-(aq) \rightleftharpoons NO(g) + 2\ H_2O(\ell) + 2\ Cl_2(g)$

The reverse reaction is favored.

EXAMPLE 19.12-WB

Write separate oxidation and reduction half-reaction equations for the redox reactions given, which occur in acidic solution. Add the half-reaction equations to produce a balanced redox equation.

$$Fe^{2+}(aq) + NO_3^-(aq) \rightarrow Fe^{3+}(aq) + NO_2(g)$$

Begin by assigning oxidation numbers to each element. Then state which element is oxidized and which element is reduced.

$Fe^{2+}(aq) + NO_3^-(aq) \rightarrow Fe^{3+}(aq) + NO_2(g)$

+2 +5/–2 +3 +4/–2

Fe is oxidized from +2 to +3. N is reduced from +5 to +4.

Write the iron(II) ion oxidation half-reaction. Balance the charge by adding electrons.

$Fe^{2+}(aq) \rightarrow Fe^{3+}(aq) + e^-$

Now (a) isolate the reduction half-reaction and (b) balance the oxygen by adding water molecules, (c) balance the hydrogen by adding hydrogen ion, and (d) balance the charge by adding electrons.

(a) $NO_3^-(aq) \rightarrow NO_2(g)$
(b) $NO_3^-(aq) \rightarrow NO_2(g) + H_2O(\ell)$
(c) $2\ H^+(aq) + NO_3^-(aq) \rightarrow NO_2(g) + H_2O(\ell)$
(d) $2\ H^+(aq) + NO_3^-(aq) + e^- \rightarrow NO_2(g) + H_2O(\ell)$

Complete the example by adding the two half-reactions. Be sure to check that the final net ionic equation is balanced in both atoms and charge.

Oxidation:	$Fe^{2+}(aq)$	$\rightarrow Fe^{3+}(aq) + e^-$
Reduction:	$2\ H^+(aq) + NO_3^-(aq) + e^-$	$\rightarrow NO_2(g) + H_2O(\ell)$
Redox:	$Fe^{2+}(aq) + 2\ H^+(aq) + NO_3^-(aq)$	$\rightarrow Fe^{3+}(aq) + NO_2(g) + H_2O(\ell)$

Check: 1 Fe, 2 H, 1 N, 3 O on each side; (2+) + 2(1+) + (1–) = 3+ on the left; 3+ on the right

EXAMPLE 19.13-WB

Write separate oxidation and reduction half-reaction equations for the redox reactions given, which occur in acidic solution. Add the half-reaction equations to produce a balanced redox equation.

$C_2O_4^{2-}(aq) + MnO_4^-(aq) \rightarrow CO_2(g) + Mn^{2+}(aq)$

Assign oxidation numbers and identify the element oxidized and the element reduced.

$C_2O_4^{2-}(aq) + MnO_4^-(aq) \rightarrow CO_2(g) + Mn^{2+}(aq)$
+3/–2 +7/–2 +4/–2 +2

C is oxidized from +3 to +4. Mn is reduced from + 7 to +2.

Complete all of the steps for the carbon oxidation half-reaction.

Partial half-reaction: $C_2O_4^{2-}(aq) \rightarrow CO_2(g)$
Balance element oxidized: $C_2O_4^{2-}(aq) \rightarrow 2\ CO_2(g)$
Balance charge with electrons: $C_2O_4^{2-}(aq) + 2\ e^- \rightarrow 2\ CO_2(g)$

Now complete the reduction half-reaction.

Partial half-reaction: $MnO_4^-(aq) \rightarrow Mn^{2+}(aq)$
Balance oxygen with water: $MnO_4^-(aq) \rightarrow Mn^{2+}(aq) + 4\ H_2O(\ell)$
Balance hydrogen with hydrogen ion: $MnO_4^-(aq) + 8\ H^+(aq) \rightarrow Mn^{2+}(aq) + 4\ H_2O(\ell)$
Balance charge with electrons: $MnO_4^-(aq) + 8\ H^+(aq) + 5\ e^- \rightarrow Mn^{2+}(aq) + 4\ H_2O(\ell)$

Add the completed half-reactions and check the final equation.

5 × Oxidation: $5\ [C_2O_4^{2-}(aq) + 2\ e^- \rightarrow 2\ CO_2(g)]$
2 × Reduction: $2\ [MnO_4^-(aq) + 8\ H^+(aq) + 5\ e^- \rightarrow Mn^{2+}(aq) + 4\ H_2O(\ell)]$

Redox: $5\ C_2O_4^{2-}(aq) + 2\ MnO_4^-(aq) + 16\ H^+(aq) \rightarrow 10\ CO_2(g) + 2\ Mn^{2+}(aq) + 8\ H_2O(\ell)$

Check: 10 C, 28 O, 2 Mn, 16 H on each side; 5(2–) + 2(1–) + 16(1+) = 4+ on the left; 2(2+) = 4+ on the right

Name: ______________________ Date: ______________________

ID: ______________________ Section: ______________________

Questions, Exercises, and Problems

Section 19.1: Electrolytic and Voltaic Cells

1. List as many things in your home as you can that are operated by voltaic cells.

2. Can a galvanic cell operate an electrolytic cell? Explain.

Section 19.2: Electron-Transfer Reactions

3. Using any example of a redox reaction, explain why such reactions are described as electron-transfer reactions.

Name: ______________________ Date: ______________________

ID: ______________________ Section: ______________________

4. Classify each of the following half-reaction equations as oxidation or reduction half-reactions:
a) $Zn \rightarrow Zn^{2+}(aq) + 2\ e^-$

b) $2\ H^+ + 2\ e^- \rightarrow H_2$

c) $Fe^{2+} \rightarrow Fe^{3+} + e^-$

d) $NO + 2\ H_2O \rightarrow NO_3^- + 4\ H^+ + 3\ e^-$

For the next two questions, classify the equation given as an oxidation or a reduction half-reaction equation.

5. Dissolving ozone in water: $O_3 + H_2O + 2\ e^- \rightarrow O_2 + 2\ OH^-$

6. Dissolving gold (Z = 79): $Au + 4\ Cl^- \rightarrow AuCl_4^- + 3\ e^-$

7. Combine the following half-reaction equations to produce a balanced redox equation:
$Cr \rightarrow Cr^{3+} + 3\ e^-$; $Cl_2 + 2\ e^- \rightarrow 2\ Cl^-$

8. The half-reactions that take place at the electrodes of an alkaline cell, widely used in portable stereos, calculators, and other devices, are
$NiOOH + H_2O + e^- \rightarrow Ni(OH)_2 + OH^-$ and $Cd + 2\ OH^- \rightarrow Cd(OH)_2 + 2\ e^-$
Which equation is for the oxidation half-reaction? Write the overall equation for the cell.

Name: ______________________________ Date: ______________________________

ID: ______________________________ Section: ______________________________

Section 19.3: Oxidation Numbers and Redox Reactions

For the next two questions, give the oxidation state of the element whose symbol is underlined.

9. $\underline{Al}^{3+}$, $\underline{S}^{2-}$, $\underline{S}O_3^{2-}$, $Na_2\underline{S}O_4$

10. $\underline{N}_2O_3$, $\underline{N}O_3^-$, $\underline{Cr}O_4^{2-}$, $NaH_2\underline{P}O_4$

In the next three questions, (1) identify the element experiencing oxidation or reduction, (2) state "oxidized" or "reduced," and (3) show the change in oxidation number. Example: $2\ Cl^- \rightarrow Cl_2 + 2\ e^-$. Chlorine oxidized from –1 to 0.

11. a) $Br_2 + 2\ e^- \rightarrow 2\ Br^-$

b) $Pb^{2+} + 2\ H_2O \rightarrow PbO_2 + 4\ H^+ + 2\ e^-$

12. a) $8H^+ + IO_4^- + 8\ e^- \rightarrow I^- + 4\ H_2O$

b) $4\ H^+ + O_2 + 4\ e^- \rightarrow 2\ H_2O$

Name: ______________________ Date: ______________

ID: ______________________ Section: ______________

13. a) $NO_2 + H_2O \rightarrow NO_3^- + 2\ H^+ + e^-$

b) $2\ Cr^{3+} + 7\ H_2O \rightarrow Cr_2O_7^{2-} + 14\ H^+ + 6\ e^-$

Section 19.4: Oxidizing Agents (Oxidizers) and Reducing Agents (Reducers)

14. Identify the oxidizing and reducing agents in $Cl_2 + 2\ Br^- \rightarrow 2\ Cl^- + Br_2$.

15. What is the oxidizing agent in the equation for the storage battery: $Pb + PbO_2 + 4\ H^+ + 2\ SO_4^{2-} \rightarrow 2\ PbSO_4 + 2\ H_2O$? What does it oxidize? Name the reducing agent and the species it reduces.

Section 19.5: Strengths of Oxidizing Agents and Reducing Agents

16. Which is the stronger reducer, Zn or Fe^{2+}? On what basis do you make your decision? What is the significance of one reducer being stronger than another?

17. Arrange the following oxidizers in order of increasing strength, that is, the weakest oxidizing agent first: Na^+, Br_2, Fe^{2+}, Cr^{2+}.

Name: ______________________ Date: ______________________

ID: ______________________ Section: ______________________

Section 19.6: Predicting Redox Reactions

In this section, write the redox equation for the redox reactants given, using Table 19.2 as a source of the required half-reactions. Then predict the direction in which the reaction will be favored at equilibrium.

18. $Br_2 + I^- \rightleftharpoons$

19. $H^+ + Br^- \rightleftharpoons$

20. $NO + H_2O + Fe^{2+} \rightleftharpoons$

Section 19.7: Redox Reactions and Acid–Base Reactions Compared

21. Explain how a strong acid is similar to a strong reducer. Explain how a strong base compares with a strong oxidizer.

Section 19.8: Writing Redox Equations (Optional)

In this section, each "equation" identifies an oxidizer and a reducer, as well as the oxidized and reduced products of the redox reaction. Write separate oxidation and reduction half-reaction equations, assuming that the reaction takes place in an acidic solution, and add them to produce a balanced redox equation.

22. $S_2O_3^{2-} + Cl_2 \rightarrow SO_4^{2-} + Cl^-$

Name: ______________________ Date: ______________________

ID: ______________________ Section: ______________________

23. $Sn + NO_3^- \rightarrow H_2SnO_3 + NO_2$

24. $C_2O_4^{2-} + MnO_4^- \rightarrow CO_2 + Mn^{2+}$

25. $Cr_2O_7^{2-} + NH_4^+ \rightarrow Cr_2O_3 + N_2$

26. $As_2O_3 + NO_3^- \rightarrow AsO_4^{3-} + NO$

General Questions

27. One of the properties of acids listed in Chapter 18 is "the ability to react with certain metals and release hydrogen." Why is the property of acids limited to certain metals? Identify two metals that do not release hydrogen from an acid and two that do.

Name: ______________________ Date: ______________________

ID: ______________________ Section: ______________________

28. Questions 22 to 26 identify reactants that engage in a redox reaction and their oxidized and reduced products. You were to write the redox equation. Nowhere among the reactants or products do you find a water molecule or a hydrogen ion. Yet, when writing the equation, you added these species. Why is this permissible?

29. It is sometimes said that in a redox reaction the oxidizing agent is reduced and the reducing agent is oxidized. Is this statement (a) always correct, (b) never correct, or (c) sometimes correct? If you select (b) or (c), given an example in which the statement is incorrect.

More Challenging Questions

30. As an example of an electrolytic cell, the text states: "Sodium chloride is electrolyzed commercially in an apparatus called the Downs cell to produce sodium and chlorine." This is a high-temperature operation; the electrolyte is molten NaCl. Write the half-reaction equations for the changes taking place at each electrode. Is the electrode at which sodium is produced the anode or the cathode?

Name: ______________________ Date: ______________________

ID: ______________________ Section: ______________________

31. Marine equipment made of iron or copper alloys and through which sea water passes sometimes has inexpensive and replaceable zinc plugs sticking into the water stream. Can you suggest a reason for this? Explain your answer.

Answers to Target Checks

1. When the nickel-cadmium (ni-cad) cell is powering the portable computer, calculator, etc., it is acting as a voltaic cell because it is a source of electrical power. When the cell is being recharged, it is acting as an electrolytic cell; the electric power from the wall outlet is the outside source that allows the ni-cad cell to recharge.

2. Oxidation is an increase in oxidation number. Reduction is a decrease in oxidation number.

3. i) a ii) b iii) d iv) c

Chapter 20

Chemical Equilibrium

☑ TARGET CHECK 20.1-WB

Consider the following statements about equilibrium systems. Identify the true statements, and rewrite the false statements so that they are true.

a) Equilibrium systems describe only physical changes.
b) Equilibrium systems describe only chemical changes.
c) Forward and reverse reaction rates are the same at equilibrium
d) The reversible changes in an equilibrium may sometimes stop, but only if both the forward and reverse reaction stop together.

☑ TARGET CHECK 20.2-WB

Consider the following statements about collision theory. Identify the true statements, and rewrite the false statements so that they are true.

a) Not all collisions result in a chemical reaction.
b) The number of effective collisions is the same as the total number of collisions.
c) A collision with too little kinetic energy does not give any product.
d) Reaction rate depends on the frequency of effective collisions.

☑ TARGET CHECK 20.3-WB

Consider the energy-reaction graph shown below for the reaction $A_2 + B_2 \rightleftharpoons 2\ AB$:

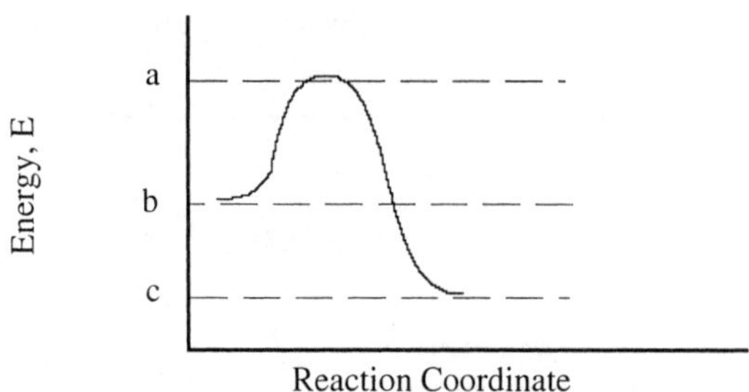

Identify the true statements concerning the graph, and rewrite the false statements so that they are true.

a) The forward reaction is exothermic.
b) The energy of the reactants is higher than the energy of the products.
c) The activation energy is the difference in energy between point **a** and point **b**.
d) An activated complex occurs at point **c**.

☑ TARGET CHECK 20.4-WB

Select the letter of the best answer choice for each of the following.

1) As temperature decreases, reaction rate decreases because:
a) the particles move too quickly for effective collisions to occur
b) the kinetic energy of the particles becomes too high for bonds to form
c) fewer particles have enough kinetic energy for an effective collision
d) the orientation of the colliding particles becomes too random for an effective collision

2) A catalyzed reaction occurs more quickly than an uncatalyzed reaction because the catalyst:
a) makes an alternative reaction pathway, with lower E_a available to the reactants
b) alters the kinetic energy of the reactants
c) changes the temperature of the mixture
d) makes the energy of reaction more negative

3) As reactant concentration falls, the reaction rate falls because the:
a) particles move too slowly for an effective collision
b) activation energy increases
c) frequency of ineffective collisions increases
d) frequency of effective collisions decreases

EXAMPLE 20.1-WB

The system $N_2(g) + O_2(g) \rightleftharpoons 2\ NO$ is allowed to come to equilibrium. Predict the direction of the Le Chatelier shift when each of the following disturbances occur. Explain your reasoning in each case.

(a) Nitrogen monoxide is added, (b) Oxygen is added, (c) Nitrogen is removed

The shift is always in the direction that tries to return the substance disturbed to its original condition. Answer and explain all three parts.

(a) Reverse, counteracting partially the increase in [NO] by consuming some of it
(b) Forward, consuming some of the added O_2
(c) Reverse, restoring some of the N_2 removed

EXAMPLE 20.2-WB

Consider the equilibrium system $2\ N_2O_5(g) \rightleftharpoons 4\ NO_2(g) + O_2(g)$. Use Le Chatelier's principle to predict the direction in which the equilibrium will shift when each of the following changes are carried out. Explain each answer.

(a) Increase the concentration of nitrogen dioxide, (b) Decrease the concentration of oxygen, (c) Decrease the concentration of dinitrogen pentoxide

(a) Reverse, consuming some of the added NO_2
(b) Forward, restoring some of the O_2 removed
(c) Reverse, restoring some of the N_2O_5 removed

EXAMPLE 20.4-WB

Predict the shift in the equilibrium system $2\ NO_2(g) \rightleftharpoons N_2O_4(g)$ that will occur because of an expansion in the volume of the container.

The increased or decreased pressure that results from the volume change will be partially offset by a change in number of molecules in the system. Decide how the volume change effects pressure, then decide whether more of fewer particles will counteract the pressure change. Those decisions will lead to the correct Le Chatelier shift.

Reverse.
The increased volume leads to a decreased pressure. The equilibrium shifts in the reverse direction to produce more particles, which partially increases the pressure.

EXAMPLE 20.5-WB

The volume of the container in which the reaction $N_2(g) + O_2(g) \rightleftharpoons 2\ NO(g)$ is decreased by a factor of 10, $V_1 = 10\ V_2$, where V_1 represents the initial volume and V_2 represents the final volume. Explain how the equilibrium will shift as a result of this volume change.

There is no shift in equilibrium position.
There are two molecules on each side of the equilibrium. The pressure of the system cannot be changed by an equilibrium shift.

EXAMPLE 20.6-WB

Consider the reaction $2\ ClF_3(g) + 2\ NH_3(g) \rightleftharpoons N_2(g) + 6\ HF(g) + Cl_2(g)$, for which $\Delta H = -1.2 \times 10^3$ kJ. Predict the direction of the Le Chatelier shift when (a) the temperature is increased and (b) when the temperature is reduced.

It is generally easier to think about temperature effects when the thermochemical equation is written with the ΔH term as part of the equation. On which side of the reaction does the "+ 1.2×10^3 kJ" term belong?

$$2\ ClF_3(g) + 2\ NH_3(g) \rightleftharpoons N_2(g) + 6\ HF(g) + Cl_2(g) + 1.2 \times 10^3\ \text{kJ}$$

The negative sign on the enthalpy-of-reaction term indicates an exothermic change, so the heat is released as a product.

Predict the shift for each temperature change.

(a) Reverse shift; (b) Forward shift
This exothermic reaction releases heat. If the temperature is increased, heat is "added," creating a stress on the right side of the equilibrium, causing an equilibrium shift in the reverse direction. The opposite is true when heat is "removed" as the temperature is reduced.

EXAMPLE 20.7-WB

All parts of this question refer to the equilibrium 148 kJ + $A_2(g) + B_3(g) \rightleftharpoons 2\ AB(g) + B(g)$.

(a) How will the equilibrium shift when some $B_3(g)$ is removed? Explain.
(b) What will happen to the equilibrium system if the container volume is increased?
(c) How is the system effected when the temperature is increased? Explain.

This question summarizes your understanding of Le Chatelier's principle. Answer all parts without further assistance.

(a) Shift in the reverse direction.
The $B_3(g)$ concentration will be partially restored to its original value by a reverse shift.
(b) Shift in the forward direction.
If volume is increased, pressure is reduced. Reduced pressure is counteracted by more molecules (3 on the right versus 2 on the left).
(c) Shift in the forward direction.
An increase in temperature can be thought of as addition of heat. The reaction responds to remove some of the added heat.

EXAMPLE 20.8-WB

Write equilibrium constant expressions for the following chemical equations:

(a) $4\ H_2(g) + CS_2(g) \rightleftharpoons CH_4(g) + 2\ H_2S(g)$

(b) $3\ O_2(g) \rightleftharpoons 2\ O_3(g)$

(c) $NO(g) \rightleftharpoons 1/2\ N_2(g) + 1/2\ O_2(g)$

(a) $K = \dfrac{[CH_4]\,[H_2S]^2}{[H_2]^4\,[CS_2]}$ (b) $K = \dfrac{[O_3]^2}{[O_2]^3}$ (c) $K = \dfrac{[N_2]^{1/2}\,[O_2]^{1/2}}{[NO]}$

EXAMPLE 20.9-WB

Write equilibrium constant expressions for the following chemical equations:

(a) $H_3PO_4(aq) + H_2O(\ell) \rightleftharpoons H_3O^+(aq) + H_2PO_4^-(aq)$

(b) $BaSO_3(s) \rightleftharpoons BaO(s) + SO_2(g)$

(c) $Al_2S_3(s) \rightleftharpoons 2\ Al^{3+}(aq) + 3\ S^{2-}(aq)$

(d) $NO_2^-(aq) + H_2O(\ell) \rightleftharpoons HNO_2(aq) + OH^-(aq)$

(e) $Mg_3(PO_4)_2(s) \rightleftharpoons 3\ Mg^{2+}(aq) + 2\ PO_4^{3-}(aq)$

(a) $K = \frac{[H_3O^+]\,[H_2PO_4^-]}{[H_3PO_4]}$ (b) $K = [SO_2]$ (c) $K = [Al^{3+}]^2\,[S^{2-}]^3$

(d) $K = \frac{[HNO_2]\,[OH^-]}{[NO_2^-]}$ (e) $K = [Mg^{2+}]^3\,[PO_4^{3-}]^2$

☑ TARGET CHECK 20.5-WB

Vinegar is a 5% by weight solution of $HC_2H_3O_2$ in H_2O. The equation $HC_2H_3O_2(aq) + H_2O(\ell) \rightleftharpoons H_3O^+(aq) + C_2H_3O_2^-(aq)$ has an equilibrium constant $K = 1.8 \times 10^{-5}$. Other than water, what substance is present in the greatest amount in vinegar? Explain.

EXAMPLE 20.10-WB

A sample of solid silver chromate, Ag_2CrO_4, is placed in pure water and allowed to come to equilibrium. An analysis of the solution indicates that the silver ion concentration is 1.3×10^{-4} M and the chromate ion concentration is 6.7×10^{-5} M. What is K_{sp} for silver chromate?

Write the equilibrium equation and equilibrium constant expression for this low-solubility solid.

$Ag_2CrO_4(s) \rightleftharpoons 2\ Ag^+(aq) + CrO_4^{2-}(aq)$ $K_{sp} = [Ag^+]^2\ [CrO_4^{2-}]$

PLAN the remainder of the problem.

GIVEN: $[Ag^+] = 1.3 \times 10^{-4}$ M; $[CrO_4^{2-}] = 6.7 \times 10^{-5}$ M *WANTED:* K_{sp}

Substitute the givens into the equilibrium constant expression and solve for K_{sp}.

$K_{sp} = [Ag^+]^2\ [CrO_4^{2-}] = (1.3 \times 10^{-4})^2\ (6.7 \times 10^{-5}) = 1.1 \times 10^{-12}$

EXAMPLE 20.11-WB

The solubility of barium phosphate is 0.0084 mg/L. Determine the value of K_{sp} for barium phosphate.

K_{sp} expressions are, by definition, expressed in terms of mol/L concentrations. Convert the given mg/L *To* mol/*L*.

GiVEN: 0.0084 mg Ba3(PO4)2/L ??Wanted: mol Ba3(PO4??2/L

Per/Path: 0.0084 mg Ba3(PO4)2/L EMBED Equation.3

g Ba3(PO4)2/L EMBED Equation.3 mol Ba3(PO4)2/L

$$0.0084 \text{ mg Ba3(PO4)2/L} \times \frac{1 \text{ g } Ba_3(PO_4)_2}{1000 \text{ mg } Ba_3(PO_4)_2} \times \frac{1 \text{ mol } Ba_3(PO_4)_2}{601.8 \text{ g } Ba_3(PO_4)_2} = 1.4 \times 10^{-8} \text{ mol } Ba_3(PO_4)_2/L$$

Now that you know the molar concentration for the ionic compound $Ba_3(PO_4)_2$, you can find the concentration in mol/L of each of the ions formed when the solid dissolved. Write the dissolving equation, K_{sp} expression, and concentration of each of the ions.

$Ba_3(PO_4)_2(s) \rightleftharpoons 3\ Ba^{2+}(aq) + 2\ PO_4^{3-}(aq)$ $K_{sp} = [Ba^{2+}]^3\ [PO_4^{3-}]^2$

$$1.4 \times 10^{-8} \text{ mol } Ba_3(PO_4)_2/L \times \frac{3 \text{ mol } Ba^{2+}}{1 \text{ mol } Ba_3(PO_4)_2} = 4.2 \times 10^{-8} \text{ M } Ba^{2+}$$

$$1.4 \times 10^{-8} \text{ mol } Ba_3(PO_4)_2/L \times \frac{2 \text{ mol } PO_4^{3-}}{1 \text{ mol } Ba_3(PO_4)_2} = 2.8 \times 10^{-8} \text{ M } PO_4^{3-}$$

You have all parts of the problem completed except for the final step, determination of K_{sp}. Calculate the final answer.

$K_{sp} = [Ba^{2+}]^3\ [PO_4^{3-}]^2 = (4.2 \times 10^{-8})^3\ (2.8 \times 10^{-8})^2 = 5.8 \times 10^{-39}$

EXAMPLE 20.12-WB

Calculate the solubility (mol/L and g/100 mL) of lead(II) iodide, which has $K_{sp} = 8.3 \times 10^{-9}$.

Write the dissolving equation and the K_{sp} expression.

$PbI_2(s) \rightleftharpoons Pb^{2+}(aq) + 2\, I^-(aq)$ $K_{sp} = [Pb^{2+}]\,[I^-]^2$

Consider the dissolving process. When one formula unit of lead(II) iodide is dissolved, what is released into solution?

One lead(II) ion and two iodide ions.
This is shown with symbols in the dissolving equation.

If you use the letter s to represent $[Pb^{2+}]$ at equilibrium, what algebraic symbol represents $[I^-]$ at equilibrium?

$[I^-] = 2s$
There are 2 iodide ions for every one lead(II) ion released into solution.

Substitute the algebraic symbols for each ion concentration into the K_{sp} expression and solve for the value of s.

$K_{sp} = [Pb^{2+}]\,[I^-]^2 = 8.3 \times 10^{-9} = (s)\,(2s)^2 = 4s^3$ $s = \sqrt[3]{\dfrac{8.3 \times 10^{-9}}{4}} = 1.3 \times 10^{-3}$

Since $s = [Pb^{2+}]$ and there is one mole of $PbI_2(s)$ for every one mole of $Pb^{2+}(aq)$ in the dissolving equation, the value of s that you just found represents the molar solubility of lead(II) iodide in solution. Add the units to the quantity, and you have the solubility in mol/L.

1.3×10^{-3} mol/L

The question also asks for solubility in g/100 mL. Use the molar solubility as a conversion factor in the setup to find the number of grams of lead(II) iodide that dissolve in 100 mL. *PLAN* and solve.

GIVEN: 100 mL *WANTED:* g PbI_2

PER/PATH: $100\text{ mL} \xrightarrow{1000\text{ mL/L}} \text{L} \xrightarrow{1.3 \times 10^{-3}\text{ mol/L}} \text{mol PbI}_2 \xrightarrow{461.0\text{ g PbI}_2\text{/mol PbI}_2} \text{g PbI}_2$

$$100\text{ mL} \times \frac{1\text{ L}}{1000\text{ mL}} \times \frac{1.3 \times 10^{-3}\text{ mol PbI}_2}{\text{L}} \times \frac{461.0\text{ g PbI}_2}{\text{mol PbI}_2} = 0.060\text{ g PbI}_2$$

The solubility is 0.060 g PbI_2/100 mL.

EXAMPLE 20.13-WB

Determine the solubility (mol/L) of silver chloride ($K_{sp} = 1.8 \times 10^{-10}$) in 0.020 M sodium chloride.

Begin with the dissolving equation and K_{sp} expression.

$AgCl(s) \rightleftharpoons Ag^+(aq) + Cl^-(aq)$ $K_{sp} = [Ag^+][Cl^-]$

The solution begins with 0.020 M sodium chloride ion already dissolved. How will that affect the dissolving equilibrium for silver chloride? In other words, what quantity do you already know in the K_{sp} expression?

$[Cl^-] = 0.020$ M

The solution is 0.020 M in sodium chloride, so it is 0.020 M in sodium ion and 0.020 M in chloride ion. The amount of chloride ion from the dissolving of the silver chloride will be insignificant compared to the 0.020 M chloride ion already in solution.

For each silver chloride formula unit that dissolves in solution, one silver ion is formed and one chloride ion is formed. The concentration of silver ion is therefore the same as the solubility of silver chloride. Complete the problem.

GIVEN: $K_{sp} = 1.8 \times 10^{-10}$, $[Cl^-] = 0.020$ M *WANTED:* $[Ag^+]$

$$[Ag^+] = \frac{K_{sp}}{[Cl^-]} = \frac{1.8 \times 10^{-10}}{0.020} = 9.0 \times 10^{-9} \text{ M}$$

EXAMPLE 20.14-WB

Determine the K_a and percentage ionization of a 0.10 M solution of a weak acid, HA, with pH = 2.65.

Begin by writing the K_a expression for the weak acid.

$$K_a = \frac{[H^+][A^-]}{[HA]}$$

You need the value for each of the three concentrations in the K_a expression.

$[H^+] = 10^{-pH} = 10^{-2.65} = 2.2 \times 10^{-3}$ M

$[A^-] = [H^+] = 2.2 \times 10^{-3}$ M

$[HA] = 0.10$ M

Determine the value of K_a.

$$K_a = \frac{[H^+][A^-]}{[HA]} = \frac{(2.2 \times 10^{-3})(2.2 \times 10^{-3})}{0.10} = 4.8 \times 10^{-5}$$

Complete the problem by calculating the percentage ionization.

$$\frac{2.2 \times 10^{-3}}{0.10} \times 100 = 2.2\% \text{ ionized}$$

EXAMPLE 20.15-WB

Find the pH of a 0.11 M solution of an acid with $K_a = 2.0 \times 10^{-5}$.

Find the hydrogen ion concentration of the acid solution.

$$[H^+] = \sqrt{K_a\,[HA]} = \sqrt{(2.0 \times 10^{-5})(0.11)} = 1.5 \times 10^{-3} \text{ M}$$

Convert from $[H^+]$ to pH.

$pH = -\log [H^+] = -\log (1.5 \times 10^{-3}) = 2.82$

EXAMPLE 20.16-WB

What is the pH of a 0.18 M solution of acetic acid ($K_a = 1.8 \times 10^{-5}$), the formula of which is often abbreviated as HAc, when the solution is also 0.10 M in acetate ion?

Take it all the way.

$HAc(aq) \rightleftharpoons H^+(aq) + Ac^-(aq)$ $\qquad K_a = \dfrac{[H^+][Ac^-]}{[HAc]}$

GIVEN: $K_a = 1.8 \times 10^{-5}$, $[HAc] = 0.18$ M, $[Ac^-] = 0.10$ M *WANTED:* pH

EQUATION: $[H^+] = K_a \times \dfrac{[HAc]}{[Ac^-]} = 1.8 \times 10^{-5} \times \dfrac{0.18}{0.10} = 3.2 \times 10^{-5}$ M

EQUATION: $pH = -\log [H^+] = -\log (3.2 \times 10^{-5}) = 4.49$

EXAMPLE 20.17-WB

If you want to make an acetic acid–acetate ion buffer with a pH of 4.90, what ratio of acetic acid to acetate ion should you use? $K_a = 1.8 \times 10^{-5}$ for acetic acid.

Make the conversion from pH to $[H^+]$, then substitute into Equation 20.9 and solve.

$[H^+] = 10^{-pH} = 10^{-4.90} = 1.3 \times 10^{-5}$

$$\frac{[HAc]}{[Ac^-]} = \frac{[H^+]}{K_a} = \frac{1.3 \times 10^{-5}}{1.8 \times 10^{-5}} = 0.87$$

EXAMPLE 20.18-WB

A researcher injects 2.00 mole of hydrogen iodide gas into a 1.00-L reaction vessel. It decomposes according to the equation $2\ HI(g) \rightleftharpoons H_2(g) + I_2(g)$. At equilibrium, 1.60 mole of HI remains. Calculate the value of the equilibrium constant at the temperature of this experiment.

Complete as much of the following table as possible, based on the given data.

	$2\ HI(g)$	$\rightleftharpoons$	$H_2(g)$	+	$I_2(g)$
mol/L at start					
mol/L change, + or –					
mol/L at equilibrium					

	$2\ HI(g)$	$\rightleftharpoons$	$H_2(g)$	+	$I_2(g)$
mol/L at start	2.00		0		0
mol/L change, + or –					
mol/L at equilibrium	1.60				

The initial concentration minus the change must be equal to the equilibrium concentration. This fact will allow you to determine the mol/L change for HI(g). You can then use the reaction stoichiometry to complete the mol/L change line for hydrogen and iodine. Insert those three values into the table above.

	$2\ HI(g)$	$\rightleftharpoons$	$H_2(g)$	+	$I_2(g)$
mol/L at start	2.00		0		0
mol/L change, + or –	– 0.40		+ 0.20		+0.20
mol/L at equilibrium	1.60				

The +0.20 mol/L change for $H_2(g)$ and $I_2(g)$ comes from the coefficients in the balanced chemical equation:

$$0.40 \text{ mol/L change for HI} \times \frac{1 \text{ mol } H_2(g) \text{ or } I_2(g)}{2 \text{ mol HI}(g)} = 0.20 \text{ mol/L change for } H_2(g) \text{ and } I_2(g).$$

Complete the problem by adding the mol/L start and mol/L change for the product species. Once you've determined the equilibrium concentrations, you can write the K expression, substitute the equilibrium concentrations, and solve for the value of K.

	$2\ HI(g)$	$\rightleftharpoons$	$H_2(g)$	+	$I_2(g)$
mol/L at start	2.00		0		0
mol/L change, + or –	– 0.40		+ 0.20		+ 0.20
mol/L at equilibrium	1.60		0.20		0.20

$$K = \frac{[H_2]\,[I_2]}{[HI]^2} = \frac{(0.20)\,(0.20)}{(1.60)^2} = 0.016$$

Name: ______________________ Date: ______________________

ID: ______________________ Section: ______________________

Questions, Exercises, and Problems

Section 20.1: The Character of an Equilibrium

1. What does it mean when an equilibrium is described as *dynamic*? Compare an equilibrium that is dynamic with one that is static.

2. Undissolved table salt is in contact with a saturated salt solution in (a) a sealed container and (b) an open beaker. Which system, if either, can reach equilibrium? Explain your answer.

3. A garden in a park has a fountain that discharges water into a pond. The pond overflows into a stream that cascades to the bottom of a small pool. The water is then pumped up into the fountain. Is this system a dynamic equilibrium? Explain.

Section 20.2: The Collision Theory of Chemical Reactions

4. Explain why a molecular collision can be sufficiently energetic to cause a reaction, yet no reaction occurs as a result of that collision.

Name: ______________________ Date: ______________________

ID: ______________________ Section: ______________________

Section 20.3: Energy Changes during a Molecular Collision

Assume heat to be the only form of reaction energy in the following questions. This makes ΔE equal to the ΔH discussed in Section 9.10.

5. Sketch an energy-reaction graph for which the answers to the first two questions in textbook Question 5 would be the opposite of what they were in that question. Include a, b, and c points on the vertical axis and use them for the algebraic expressions of ΔE and the activation energy.

6. Assuming the reaction described by Figure 20.2 to be reversible, compare the signs of the activation energies for the forward and reverse reactions. Which is positive, which is negative, or are they the same, and if so, are they positive or negative?

7. What is an "activated complex"? Why is it that we cannot list the physical properties of the species represented as an activated complex?

Section 20.4: Conditions that Affect the Rate of a Chemical Reaction

8. State the effect of a temperature increase and a temperature decrease on the rate of a chemical reaction. Explain each effect.

Name: ______________________ Date: ______________________

ID: ______________________ Section: ______________________

9. Suppose that two substances are brought together under conditions that cause them to react and reach equilibrium. Suppose that in another vessel the same substances and a catalyst are brought together, and again equilibrium is reached. How are the processes alike, and how are they different?

10. For the hypothetical reaction $A + B \rightarrow C$, what will happen to the reaction rate if the concentration of A is increases *and* the concentration of B is decreased? Explain.

Section 20.5: The Development of a Chemical Equilibrium

If nitrogen and hydrogen are brought together at the proper temperature and pressure, they will react until they reach equilibrium: $N_2(g) + 3\ H_2(g) \rightleftharpoons 2\ NH_3(g)$*. Answer the following questions with regard to the establishment of that equilibrium.*

11. When will the reverse reaction rate be at a maximum: at the start of the reaction, after equilibrium has been reached, after equilibrium has been reached, or at some point in between?

12. On a single set of coordinate axes, sketch graphs of the forward reaction rate versus time and the reverse reaction rate versus time from the moment the reactants are mixed to a point beyond the establishment of equilibrium.

Name: ______________________ Date: ______________________

ID: ______________________ Section: ______________________

Section 20.6: Le Chatelier's Principle

13. If the system $2\ SO_2(g) + O_2(g) \rightleftharpoons 2\ SO_3(g)$ is at equilibrium and the concentration of O_2 is reduced, predict the direction in which the equilibrium will shift. Justify or explain your prediction.

14. If additional oxygen is pumped into the equilibrium system $4\ NH_3(g) + 5\ O_2(g) \rightleftharpoons 4\ NO(g) + 6\ H_2O(g)$, in which direction will the reaction shift? Justify your answer.

15. Predict the direction of the shift for the equilibrium $Cu(NH_3)_4^{2+}(aq) \rightleftharpoons Cu^{2+}(aq) + 4\ NH_3(aq)$ if the concentration of ammonia were reduced. Explain your prediction.

16. A container holding the equilibrium $4\ H_2(g) + CS_2(g) \rightleftharpoons CH_4(g) + 2\ H_2S(g)$ is enlarged. Predict the direction of the Le Chatelier shift. Explain.

17. In what direction will $CO(g) + H_2O(g) \rightleftharpoons CO_2(g) + H_2(g)$ shift as a result of a reduction in volume? Explain.

Name: ______________________ Date: ______________________

ID: ______________________ Section: ______________________

18. Which direction of the equilibrium $2\ NO_2(g) \rightleftharpoons N_2O_4(g) + 59.0\ kJ$ will be favored if the system is cooled? Explain.

19. If your purpose were to increase the yield of SO_3 in the equilibrium $SO_2(g) + NO_2(g) \rightleftharpoons SO_3(g) + NO(g) + 41.8\ kJ$, would you use the highest or lowest operating temperature possible? Explain.

20. The solubility of calcium hydroxide is low; it reaches about 2.4×10^{-2} M at saturation. In acid solutions, with many H^+ ions present, calcium hydroxide is quite soluble. Explain this fact in terms of Le Chatelier's Principle. (*Hint:* Recall what you know of reactions in which molecular products are formed.)

Section 20.7: The Equilibrium Constant

For each equilibrium equation shown, write the equilibrium constant expression.

21. $CO(g) + H_2O(g) \rightleftharpoons CO_2(g) + H_2(g)$

22. $C(s) + H_2O(g) \rightleftharpoons CO(g) + H_2(g)$

Name: ______________________ Date: ______________________

ID: ______________________ Section: ______________________

23. $Zn_3(PO_4)_2(s) \rightleftharpoons 3\ Zn^{2+}(aq) + 2\ PO_4^{3-}(aq)$

24. $HNO_2(aq) + H_2O(\ell) \rightleftharpoons H_3O^+(aq) + NO_2^-(aq)$

25. $Cu(NH_3)_4^{2+}(aq) \rightleftharpoons Cu^{2+}(aq) + 4\ NH_3(aq)$

26. The equilibrium between nitrogen monoxide, oxygen, and nitrogen dioxide may be expressed in the equation $2\ NO(g) + O_2(g) \rightleftharpoons 2\ NO_2(g)$. Write the equilibrium constant expression for this equation. Then express the same equilibrium in at least two other ways, and write the equilibrium constant expression for each. Are the constants numerically equal? Cite some evidence to support your answer.

Name: ______________________ Date: ______________________

ID: ______________________ Section: ______________________

Section 20.8: The Significance of the Value of K

27. If sodium cyanide solution is added to silver nitrate solution, the following equilibrium will be reached: $Ag^+(aq) + 2\ CN^-(aq) \rightleftharpoons Ag(CN)_2^-(aq)$. For this equilibrium $K = 5.6 \times 10^{18}$. In which direction is the equilibrium favored? Justify your answer.

28. A certain equilibrium has a very small equilibrium constant. In which direction, forward or reverse, is the equilibrium favored? Explain.

29. In Chapter 17 we discussed how to identify major and minor species and how to write net ionic equations. These skills are based on the solubility of ionic compounds, the strengths of acids, and the stability of certain ion combinations. Use these ideas to predict the favored direction of each equilibrium given. In each case state whether you expect the equilibrium concentration to be large or small.

a) $H_2SO_3(aq) \rightleftharpoons H_2O(\ell) + SO_2(aq)$

b) $H^+(aq) + C_2H_3O_2^-(aq) \rightleftharpoons HC_2H_3O_2(aq)$

Name: ______________________ Date: ______________________

ID: ______________________ Section: ______________________

Section 20.9: Equilibrium Calculations (Optional)

30. $Co(OH)_2$ dissolves in water to the extent of 3.7×10^{-6} mol/L. Find its K_{sp}.

Answer:

31. If 250 mL of water will dissolve only 8.7 mg of silver carbonate, what is the K_{sp} of Ag_2CO_3?

Answer:

32. Find the moles per liter and grams per 100 mL solubility of silver iodate, $AgIO_3$, if its $K_{sp} = 2.0 \times 10^{-8}$.

Answer:

Name: ______________________ Date: ______________________

ID: ______________________ Section: ______________________

33. Find the solubility (mol/L) of $Mn(OH)_2$ if its $K_{sp} = 1.0 \times 10^{-13}$.

Answer: [ ]

34. How many grams of calcium oxalate will dissolve in 2.5×10^2 mL of 0.22 M $Na_2C_2O_4$ if $K_{sp} = 2.4 \times 10^{-9}$ for CaC_2O_4?

Answer: [ ]

35. The pH of 0.22 M $HC_4H_5O_3$ (acetoacetic acid) is 2.12. Find its K_a and percent ionization.

Answer: [ ]
[ ]

Name: ____________________ Date: ____________________

ID: ____________________ Section: ____________________

36. Find the pH of 0.35 M $HC_2H_3O_2$ ($K_a = 1.8 \times 10^{-5}$).

Answer: [ ]

37. A student dissolves 24.0 g of sodium acetate, $NaC_2H_3O_2$, in 5.00×10^2 mL of 0.12 M $HC_2H_3O_2$ ($K_a = 1.8 \times 10^{-5}$). Calculate the pH of the solution.

Answer: [ ]

38. Find the ratio $[HC_2H_3O_2]/[C_2H_3O_2^-]$ that will yield a buffer in which pH = 4.25 ($K_a = 1.8 \times 10^{-5}$).

Answer: [ ]

Name: ______________________ Date: ______________________

ID: ______________________ Section: ______________________

39. A student introduces 0.351 mol of CO and 1.340 mol of Cl_2 into a reaction chamber having a volume of 3.00 L. When equilibrium is reached according to the equation $CO(g) + Cl_2(g) \rightleftharpoons COCl_2(g)$, there are 1.050 mol of Cl_2 in the chamber. Calculate K.

Answer: ______________________

General Questions

40. At Time 1 two molecules are about to collide. At Time 2 they are in the process of colliding, and their form is that of the activated complex. Compare the sum of their kinetic energies at Time 1 with the kinetic energy of the activated complex at Time 2. Explain your conclusions.

Name: ______________________ Date: ______________________

ID: ______________________ Section: ______________________

More Challenging Questions

41. The solubility of calcium hydroxide is low enough to be listed as "insoluble" in Tables 17.3 and 17.4, but it is much more soluble than most of the other ionic compounds that are similarly classified. Its K_{sp} is 5.5×10^{-6}.
a) Write the equation for the equilibrium to which the K_{sp} is related.
b) If you had such an equilibrium, name at least two substances or general classes of substances that might be added to (1) reduce the solubility of $Ca(OH)_2$ and (2) increase its solubility. Justify your choices.
c) Without adding a calcium or hydroxide ion, name a substance or class of substances that would, if added, (1) increase $[OH^-]$ and (2) reduce $[OH^-]$. Justify your choices.

Name: ______________________ Date: ______________________

ID: ______________________ Section: ______________________

42. The table below lists several "disturbances" that may or may not produce a Le Chatelier shift in the equilibrium $4\ NH_3(g) + 7\ O_2(g) \rightleftharpoons 6\ H_2O(g) + 4\ NO_2(g)$ + energy. If the disturbance is an immediate change in the concentration of any species in the equilibrium, place in the concentration column of that substance an *I* if the change is an increase or a *D* if it is a decrease. If a shift will result, place *F* in the shift column if the shift is in the forward direction or *R* if it is in the reverse direction. Then determine what will happen to the concentrations of the other species because of the shift, and insert *I* or *D* for increase or decrease. If there is no Le Chatelier shift, write *None* in the Shift column and leave the other columns blank.

Disturbance	Shift	$[NH_3]$	$[O_2]$	$[H_2O]$	$[NO_2]$
Add NO_2					
Reduce temperature					
Add N_2					
Remove NH_3					
Add a catalyst					

43. Some systems at equilibrium are exothermic, and some are endothermic. Is this statement always true, sometimes true, or never true. Explain your answer.

44. An all-gas-phase equilibrium can be reached between sulfur dioxide and oxygen on one side of the equation and sulfur trioxide on the other. Write two equilibrium constant expressions for this equilibrium, one of which has a value greater than 1 and one with a value less than 1. It is not necessary to say which is which.

Name: ______________________ Date: ______________________

ID: ______________________ Section: ______________________

45. "Hard" water has a high concentration of calcium and magnesium ions. Focusing on the calcium ion, a common home-water-softening process is based on a reversible chemical change that can be expressed by $Na_2Ze(s) + Ca^{2+}(aq) \rightleftharpoons CaZe(s) + 2\ Na^+(aq)$. Na_2Ze represents a solid resin that is like an ionic compound between sodium ions and *zeolite ions,* a complex arrangement of silicate and aluminate groups; CaZe is the corresponding calcium compound. When, during the day, is this system most likely to reach equilibrium? Why doesn't it reach equilibrium and stay there? Periodically it is necessary to "recharge" the water softener by running salt water, NaCl(aq), through it. Why is this necessary? What concept discussed in this chapter does the recharging process illustrate?

Answers to Target Checks

1. a and b) Equilibrium systems describe both physical and chemical changes.
c) True
d) The reversible changes in an equilibrium occur continuously, never stopping.

2. a) True
b) The number of effective collisions is far less than the total number of collisions.
c) True
d) True

3. a) True
b) True
c) True for the forward reaction, false for the reverse. The activation energy for the reverse reaction is the difference in energy between point **a** and point **c**.
d) An activated complex occurs at point **a**.

4. 1) c 2) a 3) d

5. $HC_2H_3O_2$. $K = \frac{[H_3O^+][C_2H_3O_2^-]}{[HC_2H_3O_2]} = 1.8 \times 10^{-5}$. The very small value of K indicates that the denominator of the equilibrium constant fraction, $[HC_2H_3O_2]$, must be relatively large, and thus $HC_2H_3O_2$ is present in the greatest amount.

Chapter 21

Nuclear Chemistry

☑ TARGET CHECK 21.1-WB

(a) Which radioactive emission is attracted by a positive charge?
(b) Why is radioactivity sometimes termed "ionizing radiation"?

☑ TARGET CHECK 21.2-WB

What instrument uses a gas-filled tube to detect radioactivity? Briefly describe how it operates.

☑ TARGET CHECK 21.3-WB

List three important sources of radiation to which people typically are exposed to in their normal, everyday lives.

EXAMPLE 21.1-WB

(a) What fraction of a radioactive sample is left after 2, 3, 4, 5, and 6 half-lives? Answer in the form of a fraction, 1/x, where x is a counting number. (b) Convert the fractions from Part (a) to decimal numbers (round to two significant figures). (c) How does the equation $R = S \times (0.5)^n$ relate to parts (a) and (b) of this question?

After 1 half-life, 1/2 of the original sample remains. After two half-lives, half of that remains, or 1/2 of 1/2, which is $\frac{1}{2} \times \frac{1}{2} = \frac{1}{4}$. The logic is the same for more half-lives. Complete part (a) by filling in the "Fraction remaining" column of the table below.

Number of half-lives (n)	Fraction remaining	Decimal fraction remaining
2		
3		
4		
5		
6		

Number of half-lives (n)	Fraction remaining	Decimal fraction remaining
2	$\frac{1}{2} \times \frac{1}{2} = \frac{1}{4}$	
3	$\frac{1}{2} \times \frac{1}{2} \times \frac{1}{2} = \frac{1}{8}$	
4	$\frac{1}{2} \times \frac{1}{2} \times \frac{1}{2} \times \frac{1}{2} = \frac{1}{16}$	
5	$\frac{1}{2} \times \frac{1}{2} \times \frac{1}{2} \times \frac{1}{2} \times \frac{1}{2} = \frac{1}{32}$	
6	$\frac{1}{2} \times \frac{1}{2} \times \frac{1}{2} \times \frac{1}{2} \times \frac{1}{2} \times \frac{1}{2} = \frac{1}{64}$	

Complete part (b) by converting the fractions to decimal numbers. Fill in the "Decimal fraction remaining" column above.

Number of half-lives (n)	Fraction remaining	Decimal fraction remaining
2	$\frac{1}{2} \times \frac{1}{2} = \frac{1}{4}$	0.25
3	$\frac{1}{2} \times \frac{1}{2} \times \frac{1}{2} = \frac{1}{8}$	0.13
4	$\frac{1}{2} \times \frac{1}{2} \times \frac{1}{2} \times \frac{1}{2} = \frac{1}{16}$	0.063
5	$\frac{1}{2} \times \frac{1}{2} \times \frac{1}{2} \times \frac{1}{2} \times \frac{1}{2} = \frac{1}{32}$	0.031
6	$\frac{1}{2} \times \frac{1}{2} \times \frac{1}{2} \times \frac{1}{2} \times \frac{1}{2} \times \frac{1}{2} = \frac{1}{64}$	0.016

Finally, explain how the equation $R = S \times (0.5)^n$ relates to what you've just done.

The decimal fraction remaining column solves the "$(0.5)^n$" part of the equation for n = 2, 3, 4, 5, and 6.

In other words, the starting quantity multiplied by $\frac{1}{2}$ to the n (number of half-lives) power gives the remaining quantity of the radioisotope after that number of half-lives has past.

EXAMPLE 21.2-WB

The half-life of $^{31}_{14}Si$ is 2.6 hours. A sample containing 0.33 mg of silicon-31 is stored for 15.6 hours. What mass of the isotope remains?

PLAN and calculate the number of half-lives in 15.6 hours.

GIVEN: 15.6 hr; 2.6 hr/half-life *WANTED:* half-lives (n)

PER/PATH: hr $\xrightarrow{\text{2.6 hr/half-life}}$ half-lives

$$15.6 \text{ hr} \times \frac{1 \text{ half - life}}{2.6 \text{ hr}} = 6 \text{ half-lives} = n$$

PLAN and calculate the final answer.

GIVEN: S = 0.33 mg ${}^{31}_{14}Si$; n = 6 half-lives *WANTED:* R, g ${}^{31}_{14}Si$ remaining

EQUATION: $R = S \times (0.5)^n = 0.33 \text{ mg } {}^{31}_{14}Si \times (0.5)^6 = 0.0052 \text{ mg } {}^{31}_{14}Si$

EXAMPLE 21.3-WB

The isotopes of hydrogen are commonly referred to by special names. ${}^2H$ is deuterium, and ${}^3H$ is tritium. A sample containing 0.50 μg of tritium is stored for 10.0 years, after which analysis reveals 0.285 μg of tritium remaining in the sample. What is the half-life of tritium?

The overall plan for this type of problem is to first find the reaction of the sample that remains, then determine the number of half-lives that pass that correlates with the remaining fraction of the original sample, and finally find the ratio of years per half-life. What fraction remains after 10.0 years?

GIVEN: S = 0.50 μg; R = 0.285 μg *WANTED:* Fraction remaining, R/S

R/S = 0.285 μg ÷ 0.50 μg = 0.57

Now look at Figure 21.5 on textbook Page 591 and find the number of half-lives that corresponds to 0.57 of the sample remaining.

0.81 half-life has passed

Your reading of the graph may have been slightly different from ours, but it should be very close to 0.8.

Complete the problem by calculating the number of years per half-life.

$$\frac{10.0\ \text{yr}}{0.81\ \text{half-life}} = 12\ \text{yr/half-life}$$

EXAMPLE 21.4-WB

A fossilized bone sample is submitted to your laboratory for radiocarbon dating. You do an analysis of the fossil and find that its carbon-14 is decaying at a rate of 11.0 disintegrations per minute per gram of carbon. Analysis of the bone of an animal that had died recently near the site at which the fossil was recovered indicated a rate of 22.4 disintegrations per minute per gram of carbon. The half-life of carbon-14 is 5.73×10^3 years. How old is the fossil?

Find the R/S ratio and then use Figure 21.5 to determine the corresponding number of half-lives.

GIVEN: R = 11.0 disintegrations/min/g C; S = 22.4 disintegrations/min/g C

WANTED: n (number of half-lives)

R/S = 11.0 disintegrations/min/g C ÷ 22.4 disintegrations/min/g C = 0.491

1.0 half-life has past according to Figure 21.5
Also note that when R/S = 1, exactly one half-life has past.

Complete the problem. How old is the fossil?

5.73×10^3 years
One half-life is 5.73×10^3 years. No calculation is necessary.

EXAMPLE 21.5-WB

Write nuclear equations for the alpha decay of (a) radium-210 (radium, Z = 88) and (b) ^{239}Pu.

"The alpha decay" of an isotope refers to the decomposition of that isotope to an alpha particle and the appropriate second product. Begin by writing the nuclear symbol for radium-210.

$^{210}_{88}Ra$

From the periodic table, the atomic number (Z) of radium, Ra, is 88. The mass number (A) of this isotope, 210, is given in its name.

The nuclear equation will have the form $^{210}_{88}Ra \rightarrow {}^{4}_{2}He + {}^{A}_{Z}Sy$. The product $^{4}_{2}He$ is an alpha particle. Your task is to find A, the mass number, Z, the atomic number, and Sy, the chemical symbol, of the second decay product. Begin by considering A. The sum of the mass numbers must be the same on both sides of the equation: 210 = 4 + x. What is the value of x?

---

x = 206

Next, find the atomic number of the second product. You can then look at a periodic table to find its chemical symbol. Complete part (a).

$^{210}_{88}Ra \rightarrow {}^{4}_{2}He + {}^{206}_{86}Rn$

88 = 2 + x, x = 86. The last element in Group 8A/1 has atomic number 86, and its symbol is Rn, radon.

Complete part (b).

$^{239}_{94}Pu \rightarrow {}^{4}_{2}He + {}^{235}_{92}U$

From the periodic table, the atomic number of the element with the symbol Pu is 94. 239 = 4 + 235 and 94 = 2 + 92. The chemical symbol for the element with atomic number 92 is U, uranium.

EXAMPLE 21.6-WB

Write the nuclear equations for the beta decay of (a) phosphorus-32 and (b) ^{240}Np.

Beta decay is a nuclear decomposition reaction in which the products are a beta particle, ${}^{0}_{-1}e$, and the appropriate second product. The fundamental principle guiding the writing of nuclear equations remains the same: the sum of the mass numbers and the sum of the atomic numbers must be the same on opposite sides of the equation. Complete the problem.

(a) ${}^{32}_{15}P \rightarrow {}^{0}_{-1}e + {}^{32}_{16}S$ (b) ${}^{240}_{93}Np \rightarrow {}^{0}_{-1}e + {}^{240}_{94}Pu$

In part (a), you can find the atomic number of phosphorus, 15, from the periodic table. Its nuclear symbol is therefore ${}^{32}_{15}P$. In order to have the product atomic numbers add to 15, 16 is added to the −1 of the beta particle. The element with atomic number 16 is sulfur, S. The logic for part (b) is similar.

☑ TARGET CHECK 21.4-WB

(a) What is the source of induced radioactivity?

(b) Circle all transuranium elements among the following:

${}^{238}_{93}Np$ ${}^{175}_{71}Yb$ ${}^{238}_{92}U$ ${}^{143}_{59}Pr$

☑ TARGET CHECK 21.5-WB

Consider the following reactions:

(i) ${}^{238}_{92}U + {}^{1}_{0}n \rightarrow {}^{239}_{92}U$

(ii) ${}^{235}_{92}U + {}^{1}_{0}n \rightarrow {}^{142}_{56}Ba + {}^{91}_{36}Kr + 3\,{}^{1}_{0}n$

(iii) ${}^{13}_{6}C \rightarrow {}^{13}_{7}N + {}^{0}_{-1}e$

(iv) ${}^{9}_{4}Be + {}^{4}_{2}He \rightarrow {}^{12}_{6}C + {}^{1}_{0}n$

(a) Which is/are (a) fission reaction(s)? Explain.
(b) Which is/are (a) chain reaction(s)? Explain.

☑ TARGET CHECK 21.6-WB

Which of the following is/are (a) fusion reaction(s)? Explain.

(i) $^{238}_{92}\text{U} + ^{1}_{0}\text{n} \rightarrow ^{239}_{92}\text{U}$

(ii) $^{6}_{3}\text{Li} + ^{1}_{0}\text{n} \rightarrow ^{3}_{1}\text{H} + ^{4}_{2}\text{He}$

(iii) $^{13}_{6}\text{C} \rightarrow ^{13}_{7}\text{N} + ^{0}_{-1}\text{e}$

(iv) $^{3}_{1}\text{H} + ^{2}_{1}\text{H} \rightarrow ^{4}_{2}\text{He} + ^{1}_{0}\text{n}$

Name: ______________________________ Date: ______________________________

ID: ______________________________ Section: ______________________________

Questions, Exercises, and Problems

Section 21.1: The Dawn of Nuclear Chemistry
Section 21.2: Radioactivity

1. What is a nuclide? How does a nuclide differ from an isotope?

2. *Decay* is a term used to describe what happens to a radioactive nucleus. What does *decay* mean in this sense?

3. Compare the three forms of radioactive emissions in terms of mass, electrical change, and penetrating power.

Section 21.3: The Detection and Measurement of Radioactivity

4. What happens, or might happen, when an emission from a radioactive substance collides with an atom or molecule? Is this harmful? Explain.

5. What is a Geiger-Müller counter (or, simply, Geiger counter)? How does it work?

Name: ______________________ Date: ______________________

ID: ______________________ Section: ______________________

6. How do Geiger and scintillation counters differ in how they tell an observer that an object is radioactive? Can either or both be used to measure radiation as well as detect it? If so, precisely what is measured?

7. How do gamma cameras and scanners record the presence of radiation?

Section 21.4: The Effects of Radiation on Living Systems

8. Identify the greatest source of background radiation for a significant portion of the world's population. Is it a problem for that portion?

Section 21.5: Half-Life

9. The radioactivity of a sample has dropped to 1/4 of its original intensity. How many half-lives have passed?

Answer: [ ]

10. What fraction of a radionuclide remains after the passage of seven half-lives?

Answer: [ ]

Name: ____________________ Date: ____________________

ID: ____________________ Section: ____________________

11. Calculate the mass of radionuclide in a sample that will be left after 33 minutes if the sample originally has 12.9 grams of that radionuclide. The half-life of the radionuclide is 11.0 minutes.

Answer: ____________________

12. One of the more hazardous radioactive isotopes in the fallout of atomic bombs is strontium-90, for which the half-life is 28 years. If 654 g $^{90}_{38}Sr$ fall on a family farm on the day a child is born in 2004, how many grams will still be on the land when the farmer's granddaughter is born in 2060? How about when the granddaughter marries on the same farm in 2080?

Answer: ____________________

13. Uranium-235, the uranium isotope used in making the first atomic bomb, is the starting point of one of the natural radioactivity series. The next isotope in the series is thorium-231. At the beginning of a test period a sample contained 9.53 grams of the thorium isotope. After 83.2 hours only 1.05 grams of the original isotope remained. What is the half-life of thorium-231?

Answer: ____________________

Name: ______________________ Date: ______________________

ID: ______________________ Section: ______________________

14. The half-life of $^{208}_{81}Tl$ is 3.1 minutes. A 84.6-gram sample is studied in the laboratory.
a) How many grams of the isotope will remain after 12 minutes?
b) In how many minutes will the mass of $^{208}_{81}Tl$ be 3.48 grams?

Answer:

15. While excavating for the foundation of a new building, a contractor uncovered human skeletons in what turned out to be a burial ground from an ancient civilization. They were taken to a nearby university and submitted to radiocarbon dating analysis. It was found that the bones emit radiation at a rate of 55% of the rate of the rate of a living organism. How many years ago did the specimen die? (Use 5.73×10^3 years as the half-life of carbon-14.)

Answer:

Name: ______________________ Date: ______________________

ID: ______________________ Section: ______________________

Section 21.6: Natural Radioactive Decay Series—Nuclear Equations

16. What happens to the nucleus of an atom that experiences an alpha decay reaction? Compare the final nuclide with the original nuclide. Does the element undergo transmutation?

17. Write nuclear equations for the beta emissions of $^{228}_{89}Ac$ and $^{212}_{83}Bi$.

18. Write nuclear equations for the alpha decay of $^{216}_{84}Po$ and $^{234}_{92}U$.

Section 21.7: Nuclear Reactions and Ordinary Chemical Reactions Compared

19. Why is it possible to speak of the "chemical properties of lead," but not the "nuclear chemical properties of lead"?

Name: ______________________ Date: ______________________

ID: ______________________ Section: ______________________

20. The radioactivity of a sample of dirt containing uranium compounds records 5000 counts per minute when measured with a Geiger counter. The sample is treated physically to isolate the uranium compound, which is then decomposed chemically into pure uranium. If you disregard any loss of radioactivity because of decay during the purification process, will the pure uranium still radiate at 5000 counts per minute, or will it be more or less than 5000? Explain your answer.

Section 21.8: Nuclear Bombardment and Induced Radioactivity

21. Distinguish between nuclear reactions that begin spontaneously and those that begin with nuclear bombardment. What is nuclear bombardment?

22. Which of the following particles can be accelerated in particle accelerators and which cannot: electrons, protons, neutrons, positrons, alpha particles? Which property of the particle(s) governed your choice?

Name: ______________________ Date: ______________________

ID: ______________________ Section: ______________________

23. Compare the atomic numbers of all elements that are naturally radioactive with the atomic numbers of elements that exhibit artificial radioactivity.

24. Look at the periodic table. Are elements in the lanthanide series of elements transuranium elements? What about elements in the actinide series?

25. Complete each nuclear bombardment equation by supplying the nuclear symbol for the missing species.

a) ${}^{44}_{20}\text{Ca} + {}^{1}_{1}\text{H} \rightarrow ? + {}^{1}_{0}\text{n}$

b) ${}^{252}_{98}\text{Cf} + {}^{10}_{5}\text{B} \rightarrow 5\ {}^{1}_{0}\text{n} + ?$

c) ${}^{106}_{46}\text{Pd} + {}^{4}_{2}\text{He} \rightarrow {}^{109}_{47}\text{Ag} + ?$

Section 21.10: Nuclear Fission
Section 21.11: Electrical Energy from Nuclear Fission
Section 21.12: Nuclear Fusion

26. How are fission reactions like fusion reactions, and how are they different?

Name: ______________________ Date: ______________________

ID: ______________________ Section: ______________________

27. What is a chain reaction? What essential feature must be present in a nuclear reaction before it can become a chain reaction?

28. Can a fusion reaction be a chain reaction? Why or why not?

29. What advantages do fusion reactions have over fission reactions as a source of nuclear power? If fusion reactions are more desirable than fission reactions, why don't we use them instead of fission reactions?

General Questions

30. A major form of fuel for nuclear reactors used to produce electrical energy is a fissionable isotope of plutonium. Plutonium is a transuranium element. Why is this element used instead of a fissionable isotope that occurs in nature?

Name: ____________________ Date: ____________________

ID: ____________________ Section: ____________________

31. A ton of high-grade coal has an energy output of about 2.5×10^7 kJ. The energy released in the fission of one mole of $^{235}_{92}U$ is about 2.0×10^{10} kJ. How may tons of coal could be replaced by one pound of uranium-235, assuming the materials and the technology were available?

Answer: ____________________

More Challenging Questions

32. Suppose you have a radionuclide, A, that goes through a two-step decay sequence, first to B and then to C, which is stable. Suppose also that the half-life from A to B is six days, and the half-life from B to C is one day. Predict by listing in declining order, greatest to smallest, the amounts of A, B, and C that will be present (a) at the end of six days and (b) at the end of twelve days. Explain your prediction.

Name: ______________________ Date: ______________________

ID: ______________________ Section: ______________________

33. A sample of pure calcium chloride is prepared in a laboratory. A small but measurable amount of the calcium in the compound is made up of calcium-47 atoms, which are beta emitters with a half-life of 4.35 days. The compound is securely stored for a week in an inert atmosphere. When it is used at the end of that period, it is no longer pure. Why? With what element would you expect it to be contaminated?

34. Compare Equation 21.7 (Page 601) with the equation you wrote to answer textbook Question 27. Which of the two reactions is more likely to contribute to a chain reaction? Explain.

Answers to Target Checks

1. (a) A beta ray
(b) When radioactive rays collide with matter, they remove electrons.

2. A Geiger counter uses a gas-filled tube to detect radioactivity. The tube contains argon at low pressure. Radiation enters the tube and ionizes the argon. The tube is designed with a positively charged wire in the center and a negatively charged case. The ions move toward the electrodes, producing a measurable electrical pulse, which is recorded quantitatively on a meter.

3. Any three of: smoking tobacco (perhaps this is not "typical" because about 2/3 of the adult population in the world does not smoke), radon exposure, cosmic rays, x-rays, nuclear medicine, consumer products.

4. (a) Induced radioactivity comes from radioisotopes that are produced in nuclear bombardment reactions.
(b) Only $^{238}_{93}Np$ is a transuranium element.

5. (a) (ii) is a fission reaction. The products of a fission reaction have smaller atomic numbers than the starting isotope. (b) (ii) is also a chain reaction. A nuclear product of the reaction, $^{1}_{0}n$, is available to be react with another $^{235}_{92}U$ nuclide.

6. Only (iv) is a fusion reaction. A product isotope, $^{4}_{2}He$, has a larger atomic number than the staring isotopes.

Chapter 22

Organic Chemistry

☑ TARGET CHECK 22.1-WB

(a) Circle all organic compounds among the following:

CO_2 CO_3^{2-} $CH_3CH_2CH_3$ CN^- HCOOH

(b) Compare the original definition of *organic chemistry* with the modern definition. Why did the definition change?

☑ TARGET CHECK 22.2-WB

How many isomers exist for the molecular formula C_2H_4ClF? Draw their Lewis diagrams.

☑ TARGET CHECK 22.3-WB

(a) What is meant when a carbon atom is said to be saturated?

(b) An alkene has ___(i)___-bonded carbon atoms; an alkane has ___(ii)___-bonded atoms; an alkane has ___(iii)___-bonded carbon atoms.

(i) ________________________________

(ii) ________________________________

(iii) ________________________________

(c) Which of the following is not a straight-chain alkane: CH_4, C_3H_6, $C_{10}H_{22}$, C_3H_8? Explain.

(d) C_3H_7— is a(n) ________________________________ group.

(e) What is the best name for the following compound?

```
     CH3  CH2CH3
      |    |
H—————C————C—————H
      |    |
     CH3  CH3
```

(f) Draw the carbon skeleton of 2,3-dimethylbutane.

(g) What is the best name for the following compound?

```
       H     CH3
       |     |
   H---C-----C---H
       |     |
   H---C-----C---H
       |     |
      CH3    H
```

(h) Draw the carbon skeleton of 1-bromo-2-methylcyclopentane.

☑ TARGET CHECK 22.4-WB

(a) For each of the following, determine whether or not it is possible for an alkene to have that molecular formula. If it is possible, write one carbon skeleton diagram for an alkene with that formula.

(i) C_5H_{12} (ii) C_4H_8 (iii) C_2H_2

(b) Name the following compound:

$$\begin{array}{cc} CH_3 & CH_2CH_3 \\ | & | \\ C & = C \\ | & | \\ H & H \end{array}$$

(c) Draw the carbon skeleton of 1-butyne.

(d) Draw structural formulas for both *cis*- and *trans*-dichloroethylene, $C_2H_2Cl_2$.

☑ TARGET CHECK 22.5-WB

(a) What is the key structure that distinguishes an aromatic hydrocarbon?

(b) Draw the structure of ethylbenzene.

☑ TARGET CHECK 22.6-WB

Draw the structures of the products of the following reactions. Name each product.

(a)

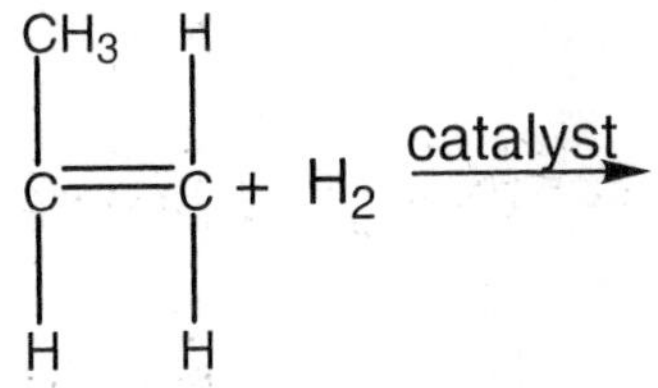

$$\begin{array}{cc} CH_3 & H \\ | & | \\ C & = C \\ | & | \\ H & H \end{array} + H_2 \xrightarrow{\text{catalyst}}$$

(b)

$$\begin{array}{ccc} H & & CH_2CH_3 \\ & C{=}C & \\ H_3CH_2C & & H \end{array} + Br_2 \longrightarrow$$

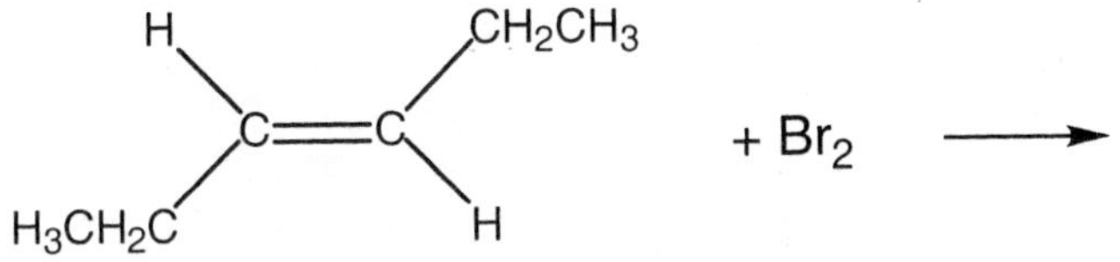

☑ TARGET CHECK 22.7-WB

(a) Write the general formula used to represent an alcohol.

(b) Write the general formula used to represent an ether.

(c) Do you expect $CH_3CH_2CH_2CH_2$–OH or $CH_3CH_2CH_2$–OH to have the greatest solubility in water? Explain your answer.

(d) Give the common name and the IUPAC name of CH_3OH.

(e) Draw the Lewis diagram of diethyl ether.

(f) Complete the following reaction:

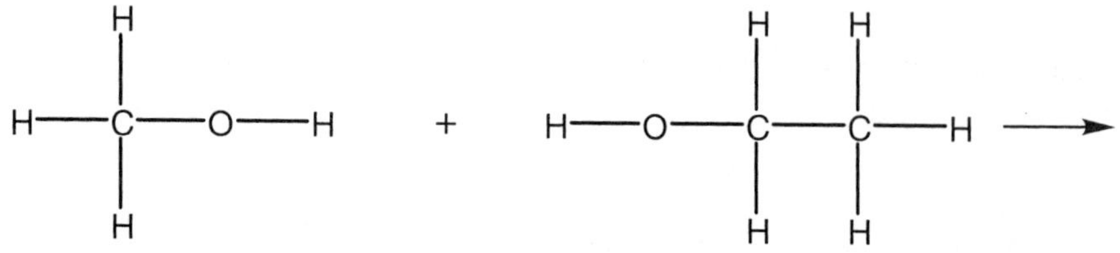

$$\mathrm{H{-}\underset{H}{\overset{H}{C}}{-}O{-}H} \quad + \quad \mathrm{H{-}O{-}\underset{H}{\overset{H}{C}}{-}\underset{H}{\overset{H}{C}}{-}H} \longrightarrow$$

☑ TARGET CHECK 22.8-WB

(a) Write the general formula used to represent an aldehyde.

(b) Write the general formula used to represent a ketone.

(c) What is the name of the following compound?

$$\mathrm{H_3C{-}CH_2{-}\overset{\overset{\displaystyle O}{\|}}{C}{-}H}$$

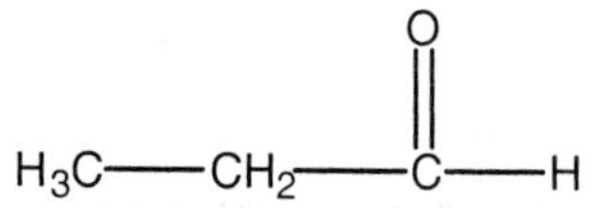

(d) What is the name of the following compound?

$$\mathrm{H_3C{-}CH_2{-}CH_2{-}\overset{\overset{\displaystyle O}{\|}}{C}{-}CH_3}$$

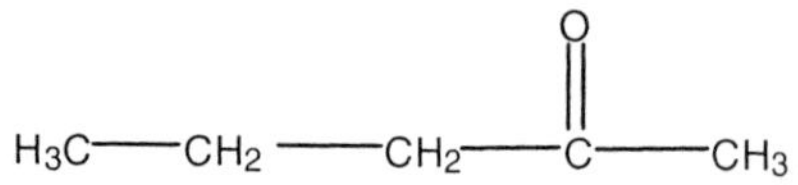

(e) Write structural diagrams to show how the ketone $CH_3–CO–CH_3$ can be prepared from an alcohol.

☑ TARGET CHECK 22.9-WB

(a) Write the general formula used to represent an ester.

(b) Write the general formula used to represent a carboxylic acid.

(c) What is the name of the following compound?

$$H_3C—CH_2—CH_2—\overset{\overset{\displaystyle O}{\|}}{C}—OH$$

(d) What is the name of the following compound?

$$H_3C—CH_2—CH_2—\overset{\overset{\displaystyle O}{\|}}{C}—O—CH_3$$

(e) Complete the following reaction:

$$H—\underset{\underset{\displaystyle H}{|}}{\overset{\overset{\displaystyle H}{|}}{C}}—\overset{\overset{\displaystyle O}{\|}}{C}—O—H \; + \; H—O—\underset{\underset{\displaystyle H}{|}}{\overset{\overset{\displaystyle H}{|}}{C}}—\underset{\underset{\displaystyle H}{|}}{\overset{\overset{\displaystyle H}{|}}{C}}—H \longrightarrow$$

☑ TARGET CHECK 22.10-WB

(a) Write the general formula used to represent an amide.

(b) Write the general formula used to represent an amine.

(c) Draw the Lewis diagram of dimethylamine.

(d) Draw the Lewis diagram of propanamide.

(e) Is dimethylamine [part (c)] a primary, secondary, or tertiary amine?

(f) Complete the following reaction:

```
       H    O
       |    ||
  H ── C ── C ── O ── H   +   H ── N ── H   ──►
       |                           |
       H                           H
```

☑ TARGET CHECK 22.11-WB

Draw the structure of 1,1-dichloroethene. The addition polymer Velon is made from this monomer. Draw three repeating units of this polymer.

☑ TARGET CHECK 22.12-WB

The structure of nylon 610 is shown below. Give the starting materials from which nylon 610 can be made by a condensation reaction.

$$—O—CH_2—CH_2—CH_2—CH_2—CH_2—CH_2—O—\overset{\overset{\displaystyle O}{\|}}{C}—CH_2—CH_2—CH_2—CH_2—CH_2—CH_2—CH_2—CH_2—\overset{\overset{\displaystyle O}{\|}}{C}—$$

Name: ______________________ Date: ______________________

ID: ______________________ Section: ______________________

Questions, Exercises, and Problems

Section 22.1: The Nature of Organic Chemistry
Section 22.2: The Molecular Structure of Organic Compounds

1. Would the cyanide ion or the carbonate ion be considered organic? What about the acetate ion?

2. What is the bond angle around a carbon atom with four single bonds? What word describes this geometry?

Section 22.3: Saturated Hydrocarbons: The Alkanes and Cycloalkanes

3. What is a hydrocarbon? Which, among the following, are hydrocarbons? CH_3OH; C_3H_4; C_8H_{10}; $CH_3CH_2CH_3$

4. Write the molecular formulas of the alkanes having 11 and 21 carbon atoms. How did you arrive at these formulas?

5. What are isomers? How does your answer to textbook Question 5 relate to this question?

Name: ______________________ Date: ______________________

ID: ______________________ Section: ______________________

6. Write the molecular formula, line formula, condensed formula, and structural formula of the normal alkane with seven carbon atoms.

7. Write the molecular formula that represents the alkyl groups having two and four carbon atoms.

8. Write the molecular formula of butane. What is the name of $C_{10}H_{22}$?

9. What is the IUPAC name of the molecule whose carbon skeleton is shown below?

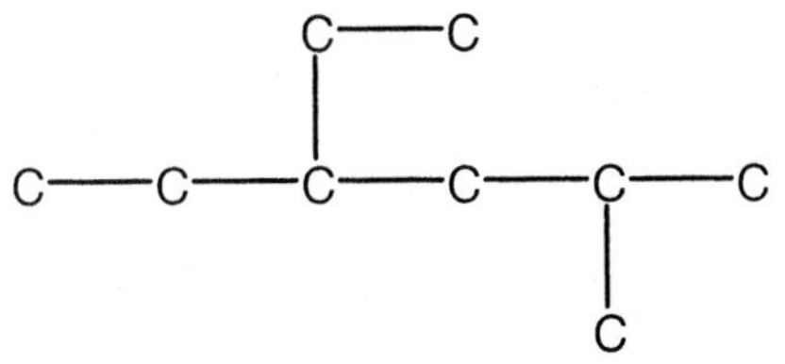

```
        C—C
        |
C—C—C—C—C—C
            |
            C
```

10. Draw the carbon skeleton of 2,3-dimethylpentane.

Name: ______________________ Date: ______________________

ID: ______________________ Section: ______________________

11. Both 1,1,1- and 1,1,2-trichloroethane are used industrially as fat and grease solvents. Draw the structural diagram of these isomers.

12. Is the general formula of a cycloalkane the same as the general formula of an alkane, C_nH_{2n+2}? Draw any structural diagrams to illustrate your answer.

13. Draw the structural diagram of cyclopentane.

14. Draw a structural diagram of 1-chloro-2-iodocyclopentane.

Name: ______________________ Date: ______________________

ID: ______________________ Section: ______________________

Questions 15–16: Write the names of the compounds whose carbon skeletons are given.

15.

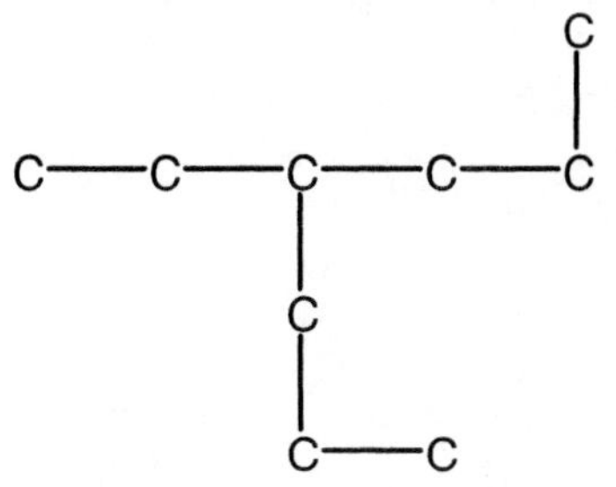

16.

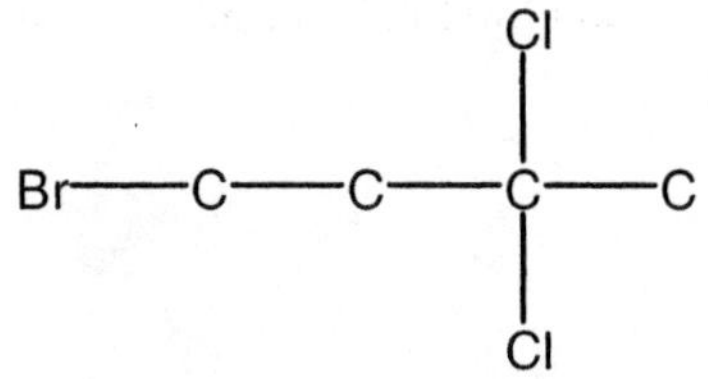

17. Draw the carbon skeleton of 2-chloro-1-iodocyclopentane.

18. Name the molecule whose carbon skeleton is drawn below.

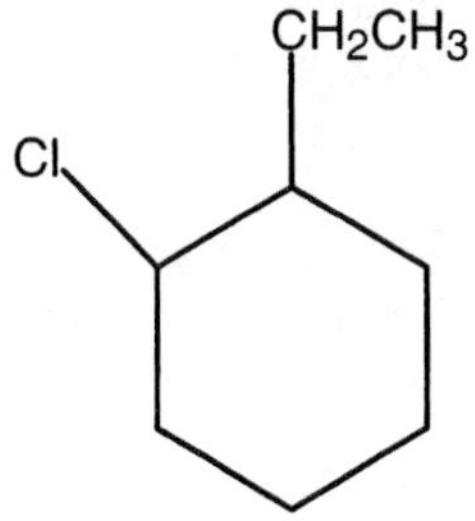

Name: ______________________ Date: ______________________

ID: ______________________ Section: ______________________

Section 22.4: Unsaturated Hydrocarbons: The Alkenes and the Alkynes

19. What is the difference in bonding and in the general molecular formula between an alkene and an alkane with the same number of carbon atoms?

20. Draw the structural formula of trichloroethene, a common dry-cleaning solvent. Why isn't the IUPAC name for this substance 1,1,2-trichloroethene?

21. Draw the structural formula and explain in words the differences between *cis*-3-heptene and *trans*-3-heptene.

22. The sex pheromone of the common house fly is *cis*-9-tricosene, where tricosene is the IUPAC name of a 23-carbon alkene. Draw the condensed formula of this molecule, marketed under the name Muscalure.

Name: ______________________ Date: ______________________

ID: ______________________ Section: ______________________

23. Give the IUPAC name of the following molecule:

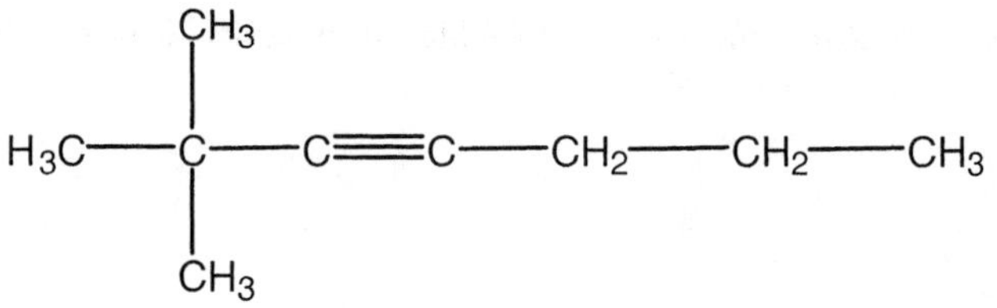

$$H_3C-C(CH_3)_2-C\equiv C-CH_2-CH_2-CH_3$$

Section 22.5: Aromatic Hydrocarbons

24. Dimethylbenzenes have the common name xylene. Draw all possible xylene isomers and give their IUPAC names.

25. Name the molecule given below.

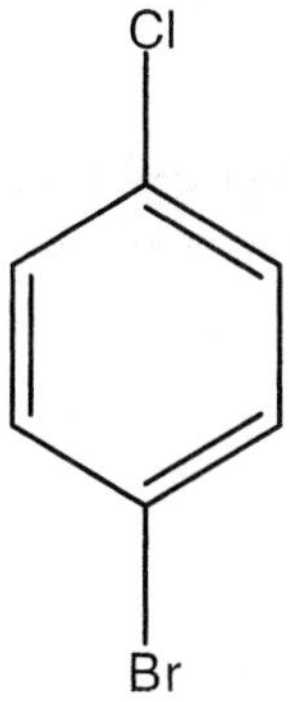

Cl

Br

Name: ______________________ Date: ______________________

ID: ______________________ Section: ______________________

Section 22.8: Chemical Reactions of Hydrocarbons

26. Draw skeletal formulas for all the possible dichloro substitution products of propane and give their IUPAC names.

27. Draw skeletal formulas for all the possible dichloro addition products of the normal butenes.

28. Write an equation for the hydrogenation of 2-butene. Does the *cis* or *trans* geometry of the butene starting material make a difference in the products obtained?

Name: ______________________ Date: ______________________

ID: ______________________ Section: ______________________

Section 22.10: Alcohols and Ethers

29. Write the Lewis structures for all possible isomers with the formula $C_4H_{10}O$. Identify them as alcohols or ethers.

30. Explain why ethers with formula $C_4H_{10}O$ have boiling points between 32°C and 39°C, whereas alcohols with the same formula have boiling points between 82°C and 118°C.

31. Write the structural formula for 2-hexanol.

32. Write the structural formula for butyl ethyl ether.

Name: ______________________ Date: ______________________

ID: ______________________ Section: ______________________

33. Write a structural equation showing how dipropyl ether might be prepared from an alcohol.

Section 22.11: Aldehydes and Ketones

34. Write structural formulas for propanal and propanone.

35. Use structural formulas to prepare acetone by oxidation of an alcohol.

Section 22.12: Carboxylic Acids and Esters

36. Write the structural formula for hexanoic acid, a wretched-smelling substance found in goat sweat.

Name: ______________________ Date: ______________________

ID: ______________________ Section: ______________________

37. Write the equation for the reaction between propanoic acid and ethanol, and name the ester formed in this reaction.

Section 22.13: Amines and Amides

38. Give structural formulas for all amines with the formula C_2H_7N and name them.

39. Classify the amines from Question 38 as primary, secondary, or tertiary.

40. Write the equation for the reaction between propanoic acid and ammonia. Name the product and give its functional group.

Name: ______________________ Date: ______________________

ID: ______________________ Section: ______________________

Section 22.15: Addition Polymers

41. Draw three repeating units of the addition polymer made from the monomer

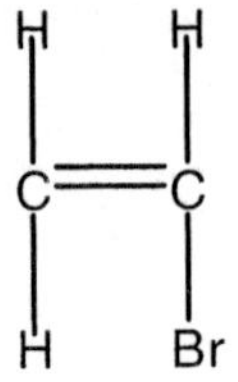

$$\begin{matrix} H & & H \\ | & & | \\ C & = & C \\ | & & | \\ H & & Br \end{matrix}$$

42. The addition polymer shown below is used for ropes, fabrics, and indoor-outdoor carpeting. Give the structure of the monomer from which this polymer was made.

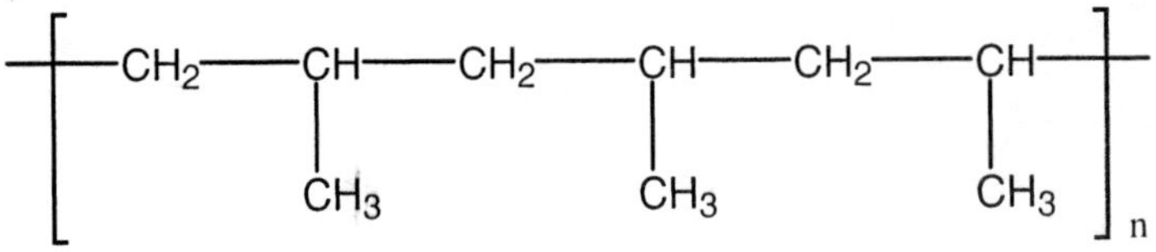

$$\left[-CH_2-\underset{\displaystyle CH_3}{\underset{|}{CH}}-CH_2-\underset{\displaystyle CH_3}{\underset{|}{CH}}-CH_2-\underset{\displaystyle CH_3}{\underset{|}{CH}}- \right]_n$$

43. Draw three repeating units of the addition polymer made from chlorotrifluoroethene.

Name: ______________________ Date: ______________________

ID: ______________________ Section: ______________________

Section 22.16: Condensation Polymers

44. Lexan is a polycarbonate ester condensation polymer that is transparent and nearly unbreakable. It is used in "bulletproof" windows (a one-inch-thick Lexan plate will stop a .38-caliber bullet fired from 12 feet), football and motorcycle helmets, and the visors in astronauts' helmets. It is made from the two monomers shown below. Draw two repeating units of this polymer.

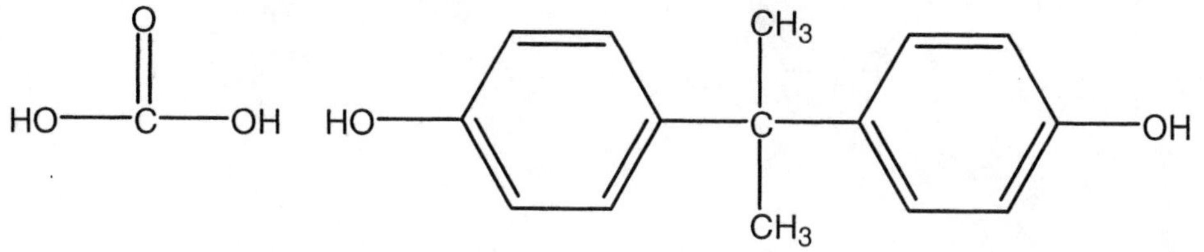

$HO-C(=O)-OH$ $HO-C_6H_4-C(CH_3)_2-C_6H_4-OH$

45. Kevlar is an aramid (see textbook Question 45). Because of its great mechanical strength, Kevlar is used in "bulletproof" clothing and in radial tires. The two monomers used to produce Kevlar are shown below. Draw two repeating units of the Kevlar polymer.

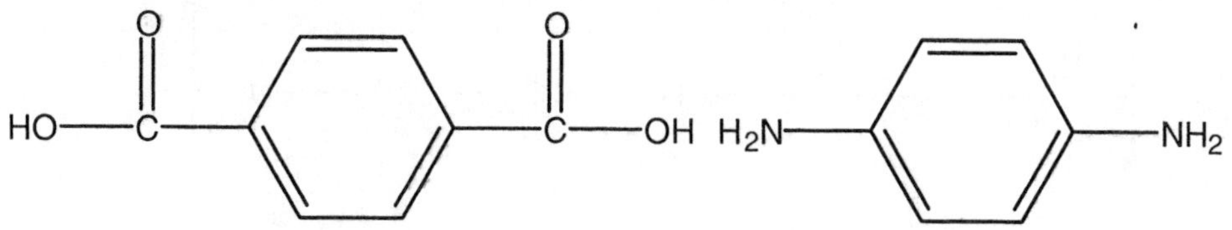

$HO-C(=O)-C_6H_4-C(=O)-OH$ $H_2N-C_6H_4-NH_2$

Name: ______________________ Date: ______________________

ID: ______________________ Section: ______________________

46. Give the monomers from which the following condensation polymer can be made.

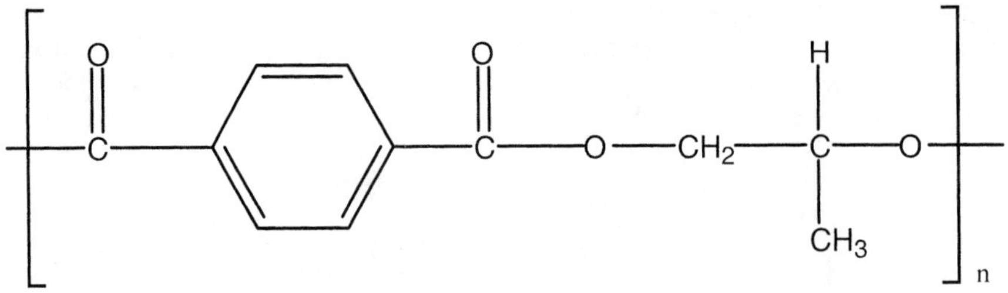

$$\left[-\overset{\overset{O}{\|}}{C}-C_6H_4-\overset{\overset{O}{\|}}{C}-O-CH_2-\underset{\underset{CH_3}{|}}{\overset{\overset{H}{|}}{C}}-O- \right]_n$$

47. A leading nylon used in Europe is nylon 6, shown below. Nylon 6 is made by polymerization of a *single, difunctional* reactant. Draw the Lewis diagram of this reactant.

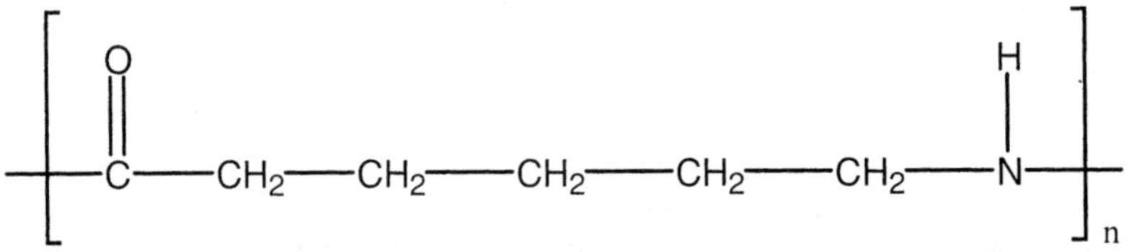

$$\left[-\overset{\overset{O}{\|}}{C}-CH_2-CH_2-CH_2-CH_2-CH_2-\overset{\overset{H}{|}}{N}- \right]_n$$

Name: ______________________ Date: ______________________

ID: ______________________ Section: ______________________

General Questions

48. Draw all isomers of C_4H_8.

49. Show that the following statement is true: "Every alcohol with two or more carbons is an isomer of at least one ether."

50. A sample of polystyrene has an average molecular mass of 1,800,000. If a single styrene molecule has the formula C_8H_8, about how many stryene molecules are in a chain of this polystyrene?

Name: ______________________ Date: ______________________

ID: ______________________ Section: ______________________

More Challenging Problems

51. Chemists often use different isotopes such as oxygen-18 to chart the path of organic reactions. If acetic acid reacts with methanol that contains only oxygen-18, show where the oxygen-18 atom exists in the ester product.

52. Write three repeating units of the addition polymer that can be made from acetylene. This material, called polyacetylene, conducts electricity because of the alternating single-bond/double-bond pattern in the main chain.

53. Amides are often made by reaction of a carboxylic acid and an amine. This reaction is both acid-catalyzed and reversible. Use these facts to explain why nylon hosiery has a very short wear life in cities where acid rain is common.

Answers to Target Checks

1. (a) $CH_3CH_2CH_3$ and HCOOH are organic. Carbon-containing compounds without hydrogen, such as CO_2, CO_3^{2-}, and CN^-, are considered to be inorganic.

   (b) Organic chemistry was originally defined as the chemistry of living organisms. The modern definition is the chemistry of carbon compounds. The definition changed because Wöhler changed an inorganic compound into an organic compound.

2. There are two isomers. The Cl and F atoms can be attached to either the same carbon or different carbons.

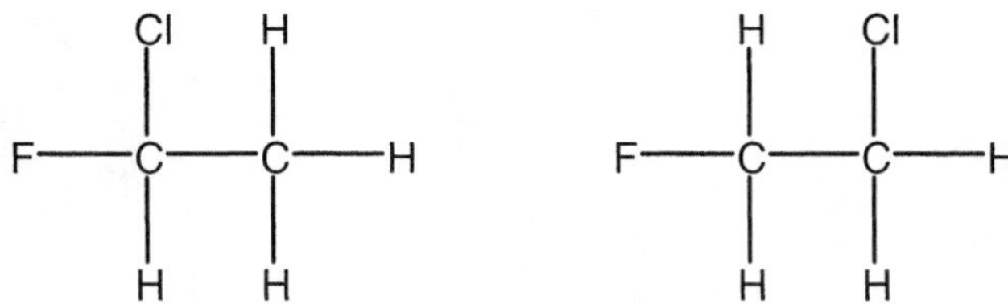

3. (a) A saturated carbon atom is bonded to the maximum number of atoms it is capable of bonding with. That maximum for carbon is four.
   (b) double; triple; single
   (c) C_3H_6 is not a straight-chain alkane. It does not have the form C_nH_{2n+2}.
   (d) Propyl
   (e) 2,3-dimethylpentane
   (f)

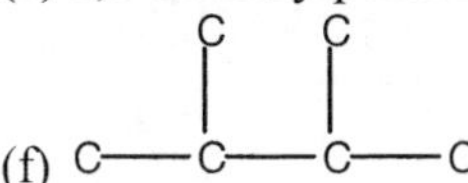

   (g) 1,3-dimethylcylcobutane
   (h)

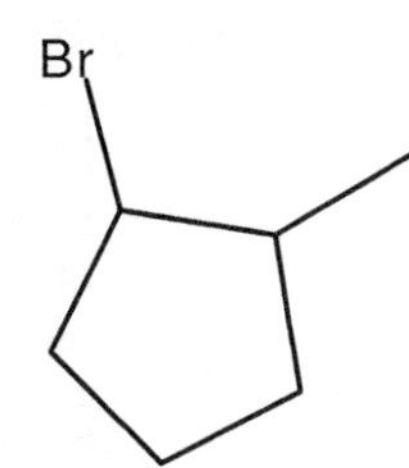

4. (a) (i) is an alkane and (iii) is an alkyne. (ii) is an alkene with two possible carbon skeletons: C = C – C – C and C – C = C – C
   (b) *cis*-2-pentene
   (c) C ≡ C – C – C
   (d)

```
            H   H                  Cl  H
            |   |                  |   |
   cis:     C = C       trans:     C = C
            |   |                  |   |
            Cl  Cl                 Cl  H
```

5. (a) Aromatic hydrocarbons contain a benzene ring.

(b)

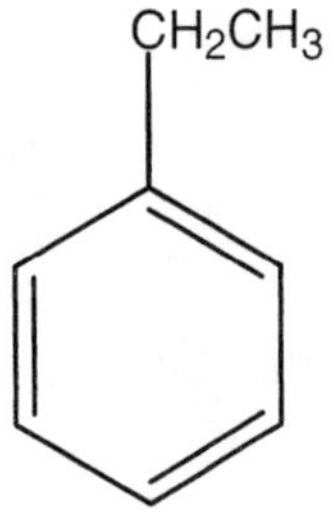

CH_2CH_3

6. (a)

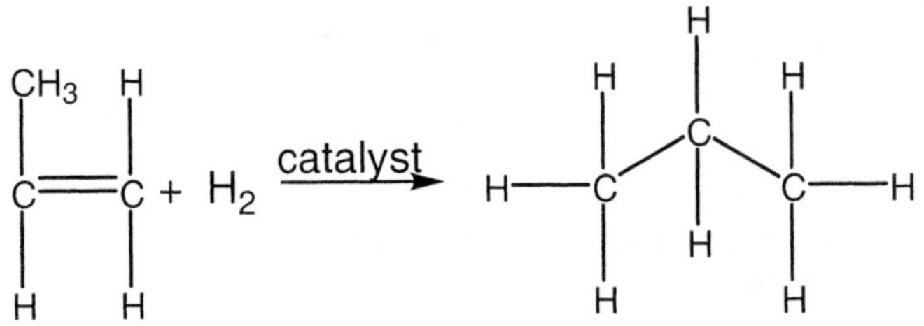

CH_3 H

C═C + H_2 $\xrightarrow{\text{catalyst}}$ H─C─C─C─H

H H

propane

(b)

H, CH_2CH_3, C═C, H_3CH_2C, H + Br_2 ⟶ H_3C─CH_2─CH(Br)─CH(Br)─CH_2─CH_3

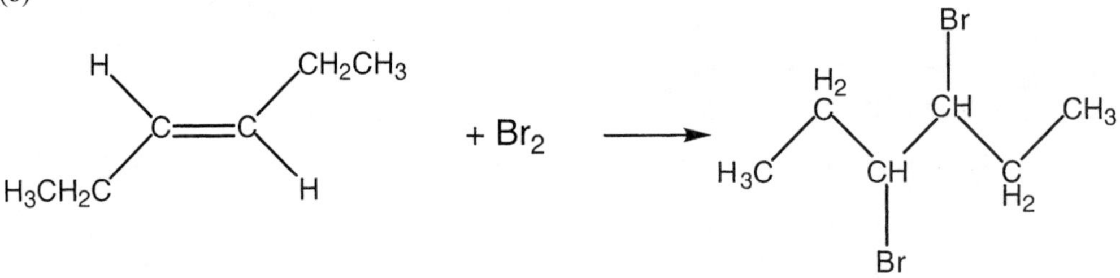

3,4-dibromohexane

7. (a) ROH

(b) ROR′

(c) One- to three-carbon alcohols, such as $CH_3CH_2CH_2$–OH, are completely soluble in water. This is primarily because of hydrogen bonding. As the carbon chain lengthens, as in $CH_3CH_2CH_2CH_2$–OH, the molecule becomes more like its parent alkane and thus less soluble in water.

(d) Common name: methyl alcohol; IUPAC name: methanol

(e)

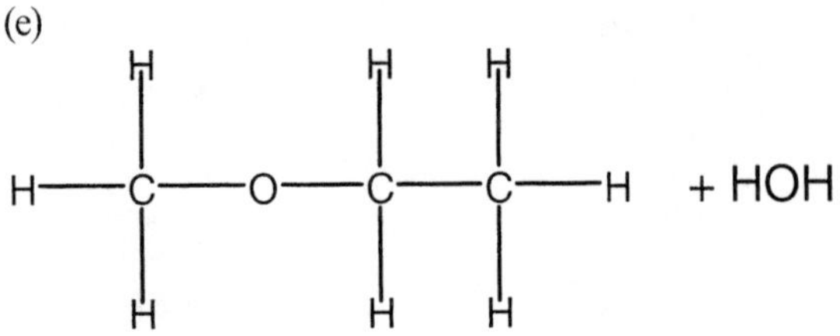

H─C─O─C─C─H + HOH

(f)

```
     H   H       H   H
     |   |       |   |
H -- C - C - O - C - C -- H
     |   |       |   |
     H   H       H   H
```

8. (a) RCHO

(b) RCOR′

(c) propanal

(d) 2-pentanone

(e)

```
        H
        |
H3C --- C --- OH  + 1/2 O2  ---->  H3C --- C == O  + H2O
        |                                  |
        CH3                                CH3
```

9. (a) RCOOR′

(b) RCOOH

(c) butanoic acid

(d) methyl butanoate

(e)

```
     H   O       H   H
     |   ||      |   |
H -- C - C - O - C - C -- H    + H2O
     |           |   |
     H           H   H
```

10. (a) $RCONR'_2$

(b) R_3N

(c)

```
H3C --- N --- CH3
        |
        H
```

(d)

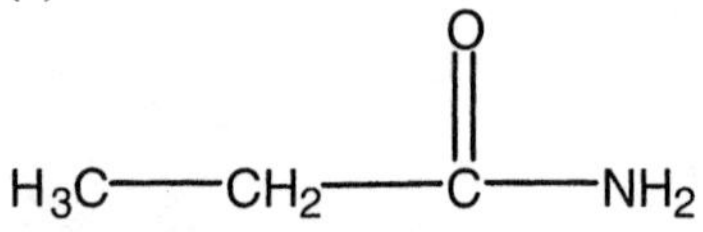

$H_3C-CH_2-C(=O)-NH_2$

(e) Secondary amine

(f)

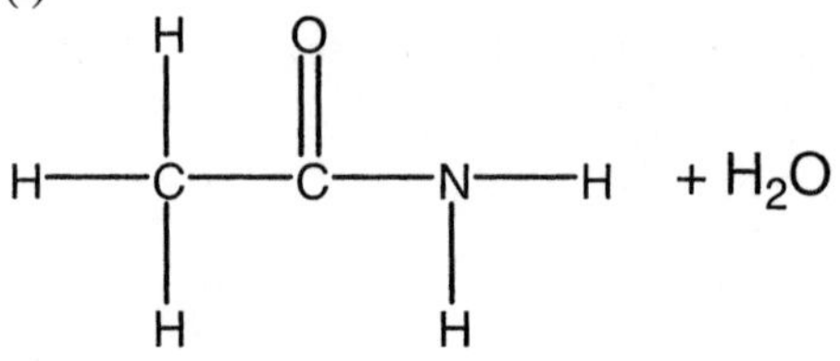

$H-CH_2-C(=O)-NH-H + H_2O$

11.

$CCl_2{=}CH_2 + CCl_2{=}CH_2 + CCl_2{=}CH_2 \longrightarrow -CCl_2-CH_2-CCl_2-CH_2-CCl_2-CH_2-$

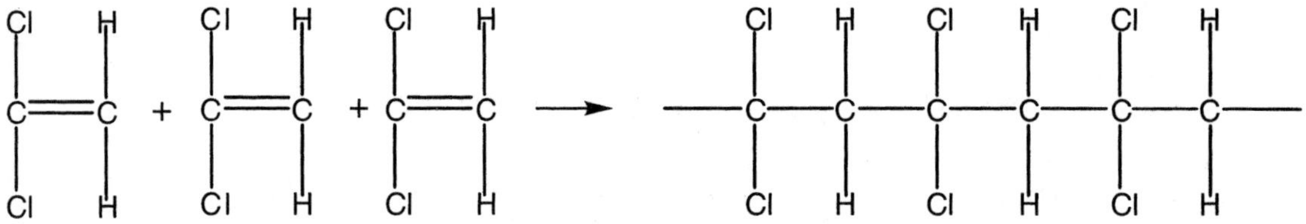

12. The dicarboxylic acid is:

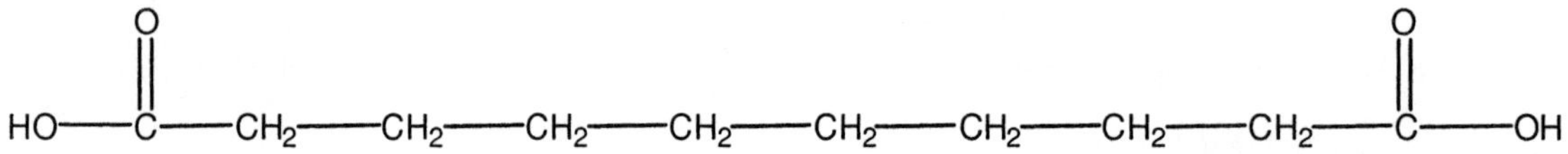

$HO-C(=O)-CH_2-CH_2-CH_2-CH_2-CH_2-CH_2-CH_2-CH_2-C(=O)-OH$

The dialcohol is:

$HO-CH_2-CH_2-CH_2-CH_2-CH_2-CH_2-OH$

Chapter 23

Biochemistry

☑ TARGET CHECK 23.1-WB

(a) Consider the peptide Asn–Glu–Gly–Phe–Asp. What is the C-terminal acid? What is the N-terminal acid?

(b) The artificial sweetener aspartame, which has the trade name NutraSweet, is the methyl ester of the dipeptide aspartylphenylalanine, abbreviated Asp–Phe or D–F. Draw the structure of aspartylphenylalanine.

(c) What does the *primary structure* of a protein refers to?

☑ TARGET CHECK 23.2-WB

(a) What is the particulate-level cause of the α-helix and β-pleated sheet secondary structures of a protein?

(b) Is hydrogen bonding in a β-pleated-sheet secondary structure at a minimum or at a maximum? Does it occur between amino acids in the same chain or in adjacent chains?

☑ TARGET CHECK 23.3-WB

(a) Define each of the following terms as they apply to enzymes: active site, inhibitor, substrate.

(b) In the lock-and-key analogy of enzyme activity, the __(i)__ is the lock and the __(ii)__ is the key.

(i) ____________________

(ii) ____________________

☑ TARGET CHECK 23.4-WB

(a) Classify each of the following as either an aldose or a ketose sugar.

H
C=O
H—C—OH
OH—C—H
H—C—OH
H—C—OH
CH_2OH

glucose

CH_2OH
C=O
HO—C—H
H—C—OH
H—C—OH
CH_2OH

fructose

H
C=O
H—C—OH
H—C—OH
H—C—OH
CH_2OH

ribose

(b) Sucrose, a glucose-fructose combination is best classified as which of the following: monosaccharide, disaccharide, polysaccharide?

(c) Melebiose is a disaccharide made from α-D-galactose and α-D-glucose monomers linked by a $\alpha(1\rightarrow6)$ bond. The monomers are shown below. Draw the structure of melebiose.

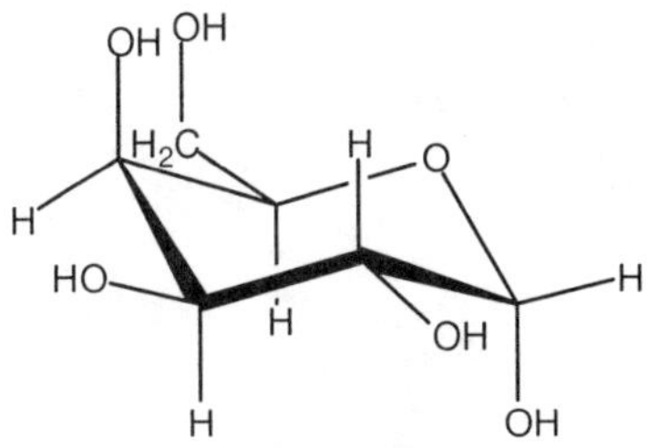

α-D-galactose

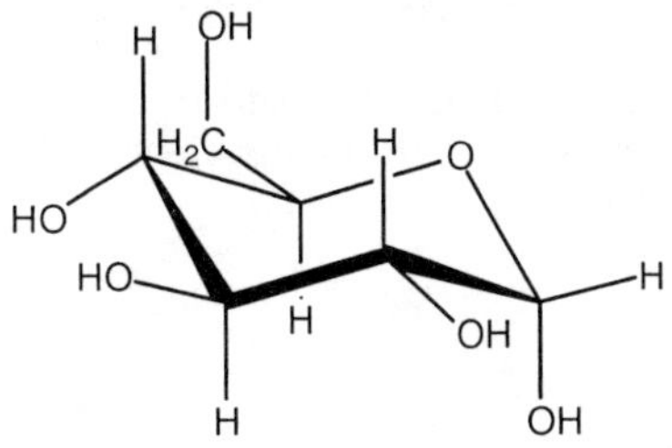

α-D-glucose

☑ TARGET CHECK 23.5-WB

Choose the letter of the best answer to each question.

(i) Which of the following statements regarding lipids is not true?
(a) Many lipids have biological roles.
(b) All lipids have the same functional groups.
(c) Lipids are soluble in nonpolar solvents.
(d) Lipids include fats, oils, waxes, and steroids.

(ii) Which of the following statements best describes the difference between fats and oils?
(a) Fats are esters of long-chain carboxylic acids; oils are esters of long-chain alcohols.
(b) Fats are esters of long-chain alcohols; oils are esters of long-chain carboxylic acids.
(c) The composition of the fatty-acid part of a fat varies; the composition of the fatty-acid part of an oil is constant.
(d) The composition of the fatty-acid part of an oil varies; the composition of the fatty-acid part of a fat is constant.
(e) Fats are triacylglycerols that are solids at room temperature; oils are triacylglycerols that are liquids are room temperature.

(iii) Which of the following best describes the structural feature(s) common to all steroid molecules?
(a) A nitrogen-containing cyclic molecule, a sugar, and one or more phosphate groups
(b) Three cyclohexane rings and one cyclopentane ring fused together
(c) A fatty acid ester with a monohydroxyl alcohol with a long carbon chain
(d) An alcohol backbone, fatty acid residues, and a phosphate group
(e) A long-chain ester of a carboxylic acid

☑ TARGET CHECK 23.6-WB

Choose the letter of the best answer to each question.

(i) What type of nucleic acid is responsible for storing genetic information and transmitting that information to the next generation during cell division?
(a) acetic acid
(b) fatty acid
(c) carboxylic acid
(d) ribonucleic acid

(e) deoxyribonucleic acid

(ii) A nucleotide is composed of:
(a) a five-carbon sugar
(b) a phosphate
(c) a nitrogen base
(d) an amino acid
(e) a, b, and c

(iii) Which of the following correctly describes the sequence in which the components of a nucleotide are ordered?
(a) monosaccharide–phosphate–nitrogen base
(b) phosphate–monosaccharide–nitrogen base
(c) nitrogen base–phosphate–monosaccharide
(d) amino acid–monosaccharide–phosphate
(e) monosaccharide–amino acid–phosphate

(iv) The structural diagram of what molecule is shown below?

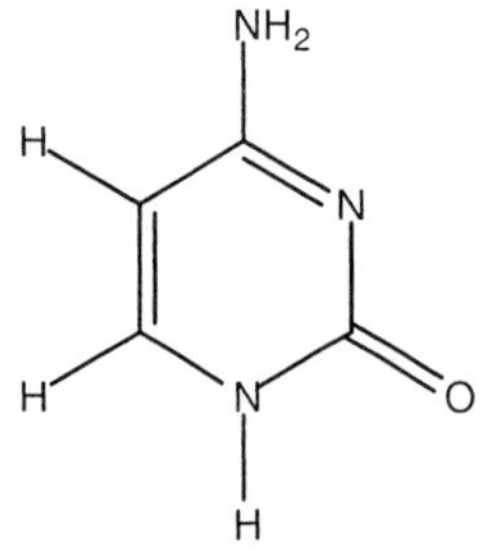

(a) adenine
(b) cytosine
(c) guanine
(d) thymine
(e) uracil

(v) Which bases in DNA pair by hydrogen bonding?
(a) cytosine and thymine
(b) cytosine and uracil
(c) adenine and guanine
(d) adenine and thymine
(e) adenine and uracil

(vi) Which of the following is the best description of the pathway from information coded in a DNA molecule to the production of a protein molecule?
(a) DNA $\xrightarrow{\text{transcription}}$ RNA $\xrightarrow{\text{translation}}$ Protein
(b) RNA $\xrightarrow{\text{transcription}}$ DNA $\xrightarrow{\text{translation}}$ Protein
(c) DNA $\xrightarrow{\text{translation}}$ RNA $\xrightarrow{\text{transcription}}$ Protein
(d) RNA $\xrightarrow{\text{translation}}$ DNA $\xrightarrow{\text{transcription}}$ Protein
(e) DNA $\xrightarrow{\text{transcription}}$ RNA $\xrightarrow{\text{replication}}$ Protein

Name: ______________________ Date: ______________________

ID: ______________________ Section: ______________________

Questions, Exercises, and Problems

Section 23.1: Amino Acids and Proteins

1. Draw the general structure of an amino acid.

2. Pick the amino acids from Table 23.1 (textbook Page 655) that have an aromatic R group.

3. Given in words the name of the tripeptides abbreviated V-T-I and I-V-T. You may use Table 23.1 if you wish. Give the name of the C-terminal amino acid in each tripeptide.

4. Write the Lewis diagram of the tripeptide abbreviated C-G-F. You may use Table 23.1 if you wish.

5. How does tertiary protein structure differ from quaternary protein structure?

Name: ______________________ Date: ______________________

ID: ______________________ Section: ______________________

6. Describe in words the hydrogen bonding that occurs in a protein having an α-helix secondary structure.

Section 23.2: Enzymes

7. To what class of biological macromolecules do enzymes belong?

8. What is an enzyme substrate?

9. How does an enzyme affect the activation energy (Section 20.3) of a reaction?

10. Many enzymes work best at temperatures near 35°C. Normal body temperature is 37.0°C. What happens to the rates of enzyme-catalyzed reactions when you run a fever?

Name: ______________________ Date: ______________________

ID: ______________________ Section: ______________________

Section 23.3: Carbohydrates

11. Give an example of an aldose sugar.

12. Examine the Lewis structures of α-glucose and β-glucose. What is the difference between these two glucose isomers?

13. To which saccharide class do the following belong? sucrose, glycogen, fructose.

14. Name the simple sugars in lactose.

15. Benedict's test is a classic test for some sugars. In this test, an aldehyde is oxidized to a carboxylic acid, and a color change occurs. Which mono- and disaccharide that you have studied would give a positive Benedict's test?

Name: ______________________ Date: ______________________

ID: ______________________ Section: ______________________

16. Maltose is a disaccharide made from two glucose molecules held together by an $\alpha(1 \rightarrow 4)$ bond. Draw the structure of maltose.

17. We have enzymes that can digest starch and turn it into energy, but we do not have enzymes that can digest cellulose. Study the structures of starch and cellulose to see how these macromolecules differ. Which bonds can our enzymes break? Which bonds can our enzymes not break?

Section 23.4: Lipids

18. What physical property do the three classes of lipids share?

19. Are oils usually obtained from animal sources or plant sources?

Name: ______________________ Date: ______________________

ID: ______________________ Section: ______________________

20. Use the letters A, B, and C to stand for three different fatty acids. Draw out all possible triacylglycerols you can make from these acids and one glycerol molecule.

21. Use the letters A and B to stand for two different fatty acids in a phospholipid. Draw out all possible phospholipids you can make from these two fatty acids.

Section 23.5: Nucleic Acids

22. In words, briefly explain the function of DNA and of RNA.

23. Draw the Lewis diagram for uracil.

24. Draw the Lewis diagrams for adenine and thymine.

Name: ______________________ Date: ______________________

ID: ______________________ Section: ______________________

25. Draw the Lewis diagram for the sugar ribose. How does this sugar differ from the sugar in textbook Question 25?

26. Draw the nucleoside adenosine, which is an adenine-ribose combination.

27. Draw the Lewis diagram for the DNA fragment that is *complementary* to the DNA fragment in textbook Question 27.

Name: ______________________________ Date: ______________________

ID: ________________________________ Section: ____________________

28. Although RNA is single stranded, the strand sometimes folds back upon itself to give a complementary portion. What would be the complementary portion of the RNA fragment in textbook Question 28?

29. Describe in words the role in protein synthesis of transfer RNA, tRNA.

Name: ______________________ Date: ______________________

ID: ______________________ Section: ______________________

General Questions

30. The tobacco mosaic virus is a small virus that has been crystallized in pure form. One complete virus has a molar mass of about 40,000,000 g/mol, of which 38,000,000 g is protein. Use the data in textbook Question 31 to determine the approximate number of amino acids in the proteins of a tobacco mosaic virus.

31. What are the monomer units for the polymers listed in textbook Question 32?

32. What element is found in DNA and RNA, but not in proteins?

More Challenging Problems

33. Rayon is an ester made by treating cotton with acetic acid. Use the cellulose structure given in Section 23.3 to draw a Lewis structure for a short rayon "molecule." (*Hint*: An "acid-base" reaction occurs randomly.)

Name: ______________________________ Date: ______________________________

ID: ______________________________ Section: ______________________________

34. RNA, unlike DNA, is a single-stranded macromolecule. Would you expect the percentage guanine in RNA to equal the percentage cytosine? Why or why not?

Answers to Target Checks

1. (a) Asp is the C-terminal acid; Asn is the N-terminal acid.

   (b)

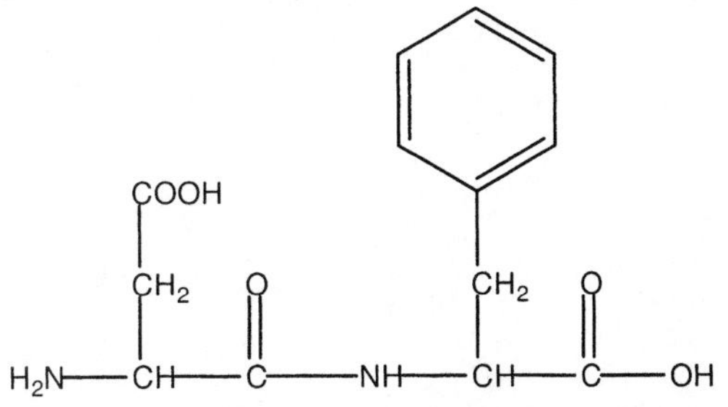

   (c) The sequence of amino acids in the protein

2. (a) Hydrogen bond formation

   (b) Maximum; in adjacent chains

3. (a) The active site is the location on an enzyme where the substrate is bound.
   An inhibitor is a compound that binds to an enzyme and interferes with its activity.
   A substrate is the reactant that an enzyme acts upon.

   (b) (i) substrate; (ii) enzyme

4. (a) Aldose sugars: glucose and ribose; ketose sugar: fructose

   (b) Disaccharide

   (c)

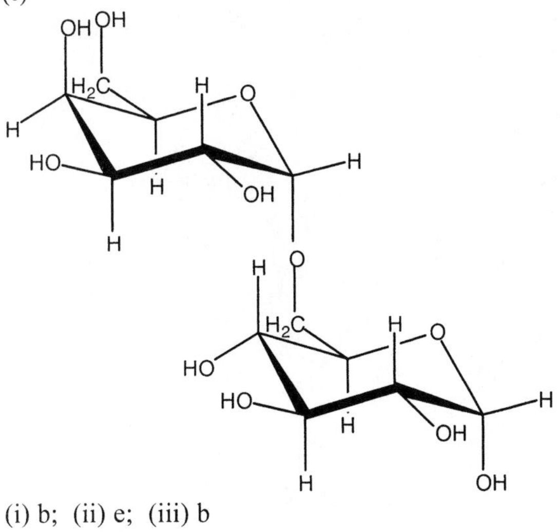

5. (i) b; (ii) e; (iii) b

6. (i) e; (ii) e; (iii) b; (iv) b; (v) d; (vi) a